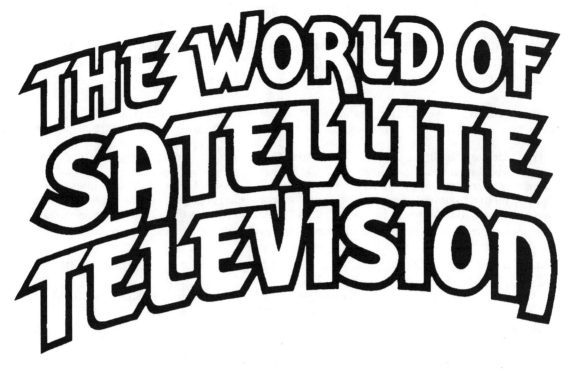

THE WORLD OF SATELLITE TELEVISION

Ninth Edition

BY MARK LONG

THE BOOK PUBLISHING COMPANY
Summertown, Tennessee

Ninth Edition—July, 1998, Eighth Edition—August, 1997, Seventh Edition—Dec. 1995, Sixth Edition—April, 1992, Fifth Edition—July, 1988, Fourth Edition—August, 1985, Third Edition—December, 1984, Second Edition—March, 1984, First Edition—January, 1983

Contributing Writers Jeffrey Keating   (Material on Installations, Troubleshooting
                                                          & MMDS Systems)
                          Peter C. Lacey (Material on Instruments for Satellite Work)
                          Doug Stevenson   (Material on SMATV Systems)
Graphic Artists
(Thailand)   *Buntung Kampenet, Jurat Landuan, Athiwat Somsri,*
                  Kittiporn Sangjun, Winai Landuan, Wonitchai Tonaboon,
              Prawood Saengsawang, and Narongrot Tepsadra
Graphic Artist
(USA)   *Peter Hoyt (material retained from the first edition).*

Special thanks   *Arthur C. Clarke, Bob Cooper and Taylor Howard.*

DEDICATION: This ninth and final edition of *The World of Satellite TV* for the millennium is dedicated to my father, Earl F. Long. During the early 1980s, he was so curious about satellite TV technology that I finally gave him a complete C-band satellite TV system. He used the system for a couple of years and then traded his 10-foot dish to someone in exchange for a sewing machine, I think. When the tiny DBS dishes started to come on line in late-1994 it started all over again. He wanted to know everything under the sun about DBS. Hopefully this book will finally lay the matter to rest. Who knows? Maybe he'll actually buy a DBS receiving system one of these days. In the meantime there's just one more thing that I want to say to dear old dad on the subject of home satellite TV: *For a guy from Michigan, you sure ask a lotta questions.*

Long, Mark.
     The world of satellite television / by Mark Long. -- 9th ed.
          p.   cm.
     Includes index.
     ISBN 1-57067-069-2 (alk. paper)
     1. Earth stations (Satellite telecommunication)-- Amateurs' manuals. 2. Direct broadcast satellite television-- Amateurs' manuals. I. Title.
     TK9962.L66 1998
     621.388'53--dc21   98-38140
                                                          CIP

Table of Contents

FOREWORD

When the history books are written for the 20th century, it is my opinion that the subject so well covered in this book, direct home reception of satellite television, will emerge as having had an effect equal to that of the automobile. It was a total surprise—even to those of us who have been involved from the beginning in the engineering and planning of space technology—that widespread use of private receiving terminals would occur so early in the 1980s. The result has been dramatic both in technical advances and price decline.

The invention of the transistor, like all non-linear advances, was simply not foreseeable. Nor was this present satellite TV revolution. Even after the invention of the transistor, none of us could imagine the marriage of materials, science, physics, the computer, and industrial technology which would give us the ability to build transistors with such short paths and so few connections that the amount of noise contributed by the transistor would be miniscule. Translated into consumer terms, yesterday's high-priced, low noise amplifier is now a one hundred dollar unit with performance beyond even the wildest dreams of a few years past.

Through such developments in solid state physics coupled with similar advances in the complexity and reliability of the satellites, we, as engineers, can see how it is technically and economically possible to make wide-band information available in the form of television, audio, and digital data to every human being on Earth.

Technology generally advances not because of planning, but because like the mountain, "it's there". Our microwave technology is now so well developed that it is no longer necessary to treat it as magic—we can ignore it in the same way that we ignore gravity. However, it cannot be ignored by our social and political planners any more than they can ignore the fact that they sit down in chairs rather than drift about the room. It is important that everyone have some understanding of at least the technical possibilities and the social implications so that we can make effective use of the technology.

Readers of this book will come to realize that they are looking at a system which has reached technical maturity and already has made a major impact on humanity. The unification of Germany and the fall of the "Iron Curtain" were due in large measure to the power of satellite technology.

With this book you have bought a ticket for a launch into the future—a rocket to much personal enjoyment—but also to a place from which you may be able to contribute to both the technology and its beneficial effects. Only those who read and then advance to having their own satellite TV system will be able to make an intelligent assessment of what this truly marvelous technology can mean to humanity.

H. Taylor Howard
Professor Emeritus,
Stanford University &
Chief Technical Officer,
Chaparral Communications

ABOUT THE AUTHOR

For the past seventeen years, Mark Long has been one of the world's leading journalists in the field of satellite TV technology. He has written more than 350 articles for magazines such as *Satellite Orbit, Satellite Retailer, Satellite TV Week, Cable and Satellite Europe, Cable and Satellite Asia, Middle East Satellites Today, Asia/Pacific Space Report, SatFACTS, Satellite Broadcasting and Tele-satellite*. Many of his articles have also appeared in the newspapers of the Los Angeles Times Syndicate.

Mark Long is the author of several books on satellite communications, including the best-selling **World of Satellite TV, The Down To Earth Guide to Satellite TV, The Inclined Orbit Satellite Tracking Guidebook** and **The Ku-band Satellite Handbook**. He also is the author of the **Satellite Series** of

educational videotapes produced by *Shelburne Films* and the creator of the new *global edition* of **The SATELLITES ON DISK Library**, an electronic reference on CD ROM disc for the commercial satellite communications industry. More than 400,000 copies of his satellite-related books have been printed. The author can be contacted at his Internet web site at http://www.mlesat.com on the worldwide web.

MLESAT.COM CD ROM

Electronic Publishing for the Satellite Professional

For Windows 95
Windows NT 4.0

Global Edition

SATELLITES ON DISK
Library

copyright 1998 Mark Long Enterprises, Inc

EXPLORING THE SATELLITE GALAXY

I n 1923, European rocket scientist Hermann Julius Oberth speculated that some day space stations orbiting the Earth would be used to provide communications services to the teeming masses of people living below. It was not until the end of World War II, however, that a comprehensive plan for such a venture was actually described in detail.

In October of 1945, a gifted science and science fiction writer proposed the extraordinary idea of using stationary satellites to beam television and other communications signals around the world. Arthur C. Clarke (*2001: A Space Odyssey, Rendezvous With Rama, The Hammer of God,* etc.) reasoned that if a satellite were positioned high enough above the Earth's equator, its orbit could be matched to the rotation of the Earth. The satellite would then appear to remain fixed in one particular spot in the sky. Because a satellite's orbital speed varies with its distance from Earth, a "geostationary" orbit is only possible

Fig. 1-1. The geostationary satellite orbit.

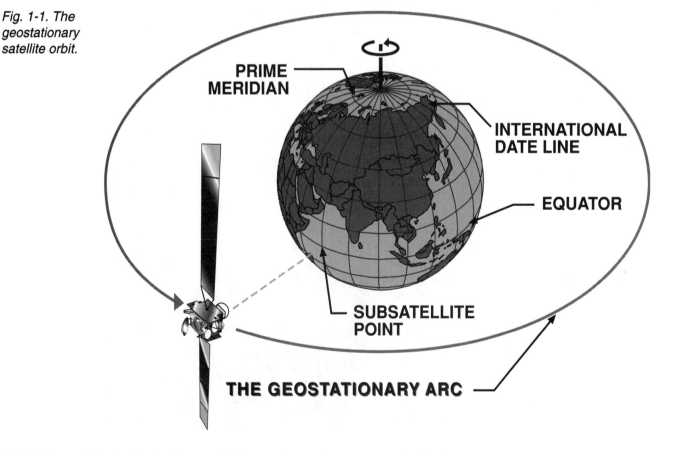

PRIME MERIDIAN

INTERNATIONAL DATE LINE

EQUATOR

SUBSATELLITE POINT

THE GEOSTATIONARY ARC

directly above the equator, in a narrow belt about 22,300 miles out. Although it took the technology a while to catch up with his simple but elegant concept, today there are hundreds of satellites taking advantage of his original thinking. In recognition of his pioneering vision, this band of outer space "real estate" is called the Clarke Orbit.

A "MODEST PROPOSAL"

Arthur C. Clarke reasoned that a single satellite located above the Earth could effectively replace thousands of local transmitting facilities at a mere fraction of their cost. He also realized that three satellites located at geostationary orbital positions over the world's major ocean regions could provide the means whereby the entire planet could be linked to a single satellite communications system.

Arthur also was the first individual to raise the possibility of direct-to-home (DTH) satellite reception as the most efficient means of supplying television and radio programming to large areas of the Earth. At first this idea was not taken very seriously in the scientific community, probably because the idea came from a writer who dabbled in the art of science fiction. But science fiction is often a springboard for new ideas and theories, some of which turn into science fact.

These days, Arthur would be the first to admit that some aspects of what he prefers to call his "modest proposal" were rendered obsolete by subsequent technological advances, all of which were totally unforeseen back in 1945. In his pioneering *Wireless World* magazine article, he imagined that future communication satellites would have to be manned platforms. Someone would have to be on station to change vacuum tubes as they burnt out. What's more,

he assumed that future satellite platforms would be extremely heavy, requiring immense rocket power to lift them into space. The development of solid-state electronic devices such as the transistor and the photo-voltaic power cell during the 1950s made it possible for engineers to dramatically reduce the size, weight, and power requirements of telecommunication satellites.

ADVENT OF THE SATELLITE AGE

On October 4, 1957, the Soviet Union launched Sputnik Zemli (Russian for "companion of the world"), the world's first artificial satellite. In the West, the satellite was simply known as Sputnik. Although it only transmitted a simple beacon that could be tracked as the satellite orbited the Earth, its successful flight had implications that challenged the rest of the world to develop the advanced technologies necessary for the production of communications satellites.

On July 10, 1962, the United States launched Telstar, the world's first communications satellite used to relay television programming. Because satellite-launching techniques were still in their

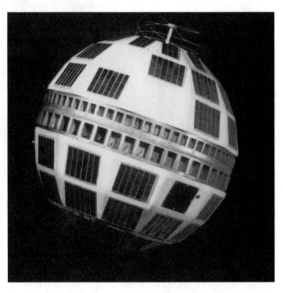

Fig. 1-2. The Telstar satellite. (Courtesy NASA.)

infancy, Telstar was not put into geostationary orbit. Its elliptical orbit had to be tracked by ground stations and thus could only provide telecommunication services for a maximum of 102 minutes per day. Telstar was responsible for providing the first live transatlantic relays of TV programming.

Launched on August 19, 1964, Syncom 3 was the first communications satellite with 24-hour availability to transmit live TV signals, some of which came from the *1964 Olympic Games* held in Tokyo, Japan. The success of Syncom 3 led to the establishment of an international cooperative charged with the commercial development of a global satellite communications system.

In 1965, the Early Bird satellite became operational as the world's first commercial geostationary satellite. It could carry 240 telephone communications circuits, or one television channel, at a time. On June 2, 1965, Early Bird introduced live television across the Atlantic Ocean. Early Bird was the first satellite to be owned and operated by the International Telecommunications Satellite Organization (INTELSAT).

In April of 1965, the Soviet Union launched its first communications satellite. Called Molniya (Russian for "lightning"), the satellite occupied an orbit that was not synchronous with the Earth's rotation. Instead, the spacecraft used an elliptical orbit that was inclined approximately 65 degrees from the Earth's equator. This allowed the satellite to relay communications traffic to locations in the farthest northern regions of the country. Within two days of launch, Molniya 1 was broadcasting black and white TV signals between Moscow and Vladivostok. In October of 1965, the satellite also conducted an experimental TV broadcast between Paris and Moscow.

First Canada, then the United States, and subsequently other countries constructed their own domestic geostationary satellite systems. Each new satellite had greater capabilities, expanding our ideas of the technologically possible. As of early 1998, there were more than 150 domestic and international communications satellites in geostationary orbit over the Earth's equator.

UPLINKS & DOWNLINKS

Each satellite is both a receiver and a transmitter. First, a ground station, also called the uplink, sends a signal to the satellite. The satellite automatically changes the signal's frequency and retransmits it back to stations on the ground. This second signal path is called the downlink. A satellite is much like a broadcasting tower 22,300 miles high, an automatic relay station that can transmit into a coverage area which encompasses up to 42.4 percent of the Earth's surface.

Each satellite has a number of redundant modules, spare components that

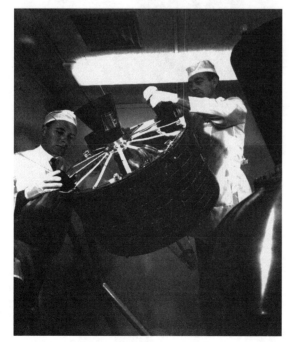

Fig. 1-3. The Syncom satellite. (Courtesy NASA.)

can be switched into operation in the event that any malfunction of the primary parts occurs. A satellite typically costs US $200 million or more for the spacecraft, insurance and launching rocket—too expensive a project to allow parts failure to disrupt the services. There aren't any repairmen who can make a service call to the Clarke Orbit. But ground control stations can remotely switch in backup facilities in case of failure.

SATELLITE TRANSPONDERS

Every communications satellite carries several channels, called transponders, which process communications traffic. Most satellites have sixteen or more transponders in operation, each capable of transmitting one or more TV signals as well as thousands of simultaneous telephone conversations. Radio networks are also present on many transponders, as well as facsimile, computer information services, and other data transmissions.

Each transponder's signal uses one of two available senses of polarization (either horizontal/ vertical or right-hand circular/left-hand circular), which are at right angles or "orthogonal" to one another. Signals of opposite polarization are received separately at the receiving earth station. Polarization allows more channels to fit into the limited range of frequencies allocated for satellite communications. Most telecommunication satellites carry eight or more transponders using one sense of polarization and eight or more additional transponders using the opposite sense of polarization. This allows each spacecraft to reuse the available satellite frequency spectrum twice.

The satellite frequency "bands" are located high above those used by all earth-bound "terrestrial" TV channels. These super-high frequencies are not affected by sunspots or any other atmospheric conditions; satellites, therefore, provide extremely reliable communications coverage 24 hours a day.

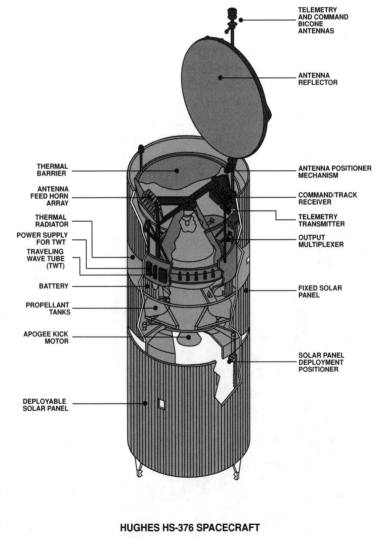

Fig. 1-4. The Hughes HS-376 is the most widely used spacecraft design in the history of satellite communications.

TELEMETRY AND COMMAND BICONE ANTENNAS

ANTENNA REFLECTOR

THERMAL BARRIER

ANTENNA FEED HORN ARRAY

THERMAL RADIATOR

POWER SUPPLY FOR TWT

TRAVELING WAVE TUBE (TWT)

BATTERY

PROPELLANT TANKS

APOGEE KICK MOTOR

DEPLOYABLE SOLAR PANEL

ANTENNA POSITIONER MECHANISM

COMMAND/TRACK RECEIVER

TELEMETRY TRANSMITTER

OUTPUT MULTIPLEXER

FIXED SOLAR PANEL

SOLAR PANEL DEPLOYMENT POSITIONER

HUGHES HS-376 SPACECRAFT

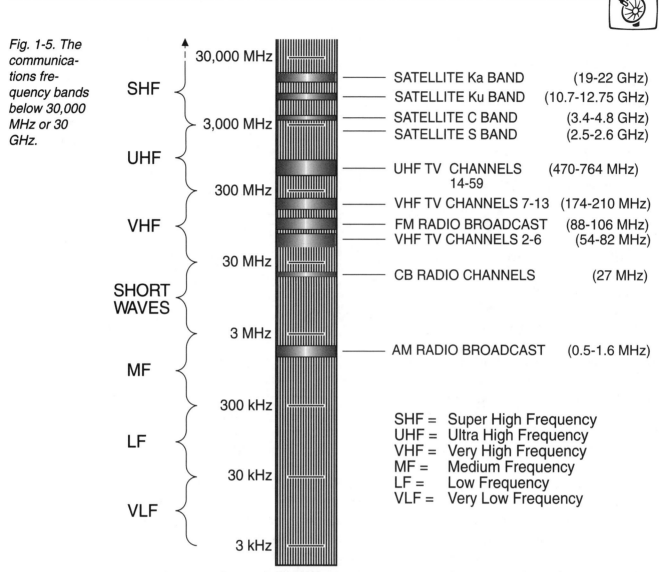

Fig. 1-5. The communications frequency bands below 30,000 MHz or 30 GHz.

SATELLITE Ka BAND	(19-22 GHz)
SATELLITE Ku BAND	(10.7-12.75 GHz)
SATELLITE C BAND	(3.4-4.8 GHz)
SATELLITE S BAND	(2.5-2.6 GHz)
UHF TV CHANNELS 14-59	(470-764 MHz)
VHF TV CHANNELS 7-13	(174-210 MHz)
FM RADIO BROADCAST	(88-106 MHz)
VHF TV CHANNELS 2-6	(54-82 MHz)
CB RADIO CHANNELS	(27 MHz)
AM RADIO BROADCAST	(0.5-1.6 MHz)

SHF = Super High Frequency
UHF = Ultra High Frequency
VHF = Very High Frequency
MF = Medium Frequency
LF = Low Frequency
VLF = Very Low Frequency

Operating at frequencies of several billion cycles per second, or Gigahertz (GHz), the region's satellites relay TV programming via two distinct communication bands. North America's high-power Direct Broadcast Satellite (DBS) services use frequencies ranging from 12.2 to 12.7 GHz. A few operators using medium-power satellites also offer direct-to-home (DTH) TV services using adjacent frequencies from 11.7 to 12.2 GHz. The entire frequency spectrum from 10.7 to 12.75 GHz is commonly referred to as the "Ku" (pronounced "Kay-You") band. Throughout the Americas, numerous other satellites are available that operate within a lower fre-

quency spectrum ranging from 3.7 to 4.2 GHz. This frequency range is known as the "C" band. Home satellite TV receiving systems are available that can receive TV signals from either or even both of these frequency bands.

Overseas, there are other satellites that carry transponders designed to transmit in the 2.5 to 2.65 GHz range that has been assigned for DTH and community TV reception applications. This frequency spectrum commonly is referred to as the "S" band. What's more, future telecommunication spacecraft are now on the drawing boards that will soon be using even higher "Ka" band frequencies ranging from 19 to 22 GHz.

FOOTPRINTS

The satellite's antenna transmits the television signal in a particular shape, called a footprint, pointed at its assigned coverage area. Each satellite generates one or more characteristic footprints, with the signal strongest at the center of each footprint's coverage area and diminishing outward from there. Those who live at locations toward the center of any footprint can receive the signals with a smaller dish antenna than those out on the edges. For example, a TV viewer in Iowa may need an antenna that is 15 inches in diameter to receive the same quality of service from a particular Ku-band satellite as an individual in Montana using an antenna which is 24 inches in diameter.

C-band satellite transponders typically transmit low-powered signals, usually in the neighborhood of sixteen to thirty watts—not much more power than a car tail light bulb uses. By the time the satellite signal reaches the ground, it is very weak indeed. That is one reason why large aperture antennas (commonly 1.8 to 3.0 meters in diameter) must be used to pull the C-band signals out of the background noise.

Ku-band operating within the 11.7 to 12.2 GHz frequency range transmit medium-powered signals, usually in the neighborhood of 50 to 100 watts, while high-powered satellites operating in the 12.2 to 12.7 GHz frequency spectrum, also referred to as DBS (for direct broadcast satellite) spacecraft, transmit 120 to 240 watts of power per Ku-band transponder. Higher power from the

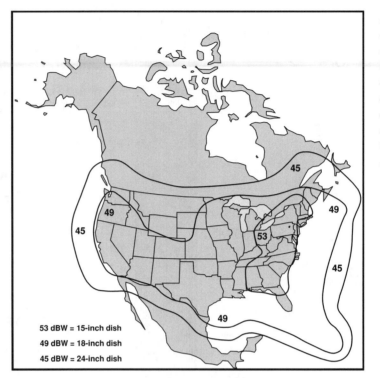

53 dBW = 15-inch dish
49 dBW = 18-inch dish
45 dBW = 24-inch dish

Fig. 1-6. A representative DBS satellite footprint map showing dish sizes required at various locations throughout North America.

satellite directly translates into smaller dishes down on the ground—as small as 45 centimeters in diameter.

THE BIRTH OF THE COMSAT

In the 1960s, every communications satellite, or "comsat," was operated exclusively by international organizations and government-owned telecommunications authorities that had the wherewithal to construct and operate space communications systems. Reception of early TV transmissions required extremely large parabolic dishes and expensive electronic components, which only these large entities could afford.

On June 25, 1967, the world was linked "live via satellite" for the very first time. Called *Our World,* the two-hour global telecast reached an estimated 600 million TV viewers in twenty-six nations. The program started out with live coverage of a baby's birth in a Mexico City hospital and then swept around the globe to show the very best that the world had to offer. In New York City,

Fig. 1-7. The Communications Technology Satellite undergoes prelaunch checkout in Hangar S at Cape Canaveral Air Force Station. (Courtesy NASA,)

Leonard Bernstein and Van Cliburn displayed their musical talents. When live coverage switched to Italy, viewers saw director Franco Zefferelli hard at work on his new movie *Romeo and Juliet.* In the "land down under," an Australian astronomer peered through a telescope at what was then the most distant known object in the universe. For many, however, the culmination of the two-hour broadcast came when the Beatles introduced a new song entitled *All You Need Is Love.* For many members of my generation, it was the perfect kick-off for 1967's now famous *Summer of Love.*

In this age of instantaneous telecommunication—where live coverage concerning the death of Diana, Princess of

Fig. 1-8. A closeup view of the Communications Technology Satellite's antenna payload. At the time of launch, CTS was the world's most powerful satellite.

Wales, can become known worldwide with in a matter of minutes—it is difficult to believe that it took two years to organize the 1967 *Our World* satellite broadcast. Two of the three satellites used for the inaugural live global broadcast hadn't even been launched when the event's first planning session was held. But two years of hard work involving a cast of thousands paid off: the "global village" had finally arrived.

THE CTS BROADCASTING PROJECT
In 1971, direct-to-home (DTH) satellite broadcasting took its first step toward becoming a reality when Canada and the United States established a joint project to develop, launch and operate an experimental platform called the Communications Technology Satellite (CTS). The main goal of the project was to develop and test the technologies required to transmit TV programming directly to individual homes and communities equipped with small aperture receiving antennas. The challenge was to design and manufacture satellite components that had never before been produced—transponders capable of operating in the Ku-band frequency spectrum as well as the high-power amplifiers needed to make the small receiving dish a practicality.

Canada's Communications Research Center (CRC) was responsible for the design and construction of the CTS spacecraft while NASA provided the high-power amplifier tubes and the launch vehicle. The European Space Agency, which joined the project in 1972, also provided several spacecraft components, including the solar arrays.

THE HERMES EXPERIMENTS
On January 17, 1976, a NASA Delta rocket successfully boosted the Communications Technology Satellite into

orbit. Following a successful launch, the spacecraft was re-named "Hermes" after the god of science and invention in Greek mythology. Hermes inaugurated the first in a series of experimental TV transmissions on May 21, 1976.

The first ever direct-to-home satellite TV broadcast took place on May 16, 1978, when Hermes transmitted a Canadian hockey game to a 60cm dish installed at the home of a Canadian embassy official in Lima, Peru. Attendees from an international conference were on hand to witness the event. In August of 1979, the satellite was relocated to a new orbital assignment over the Pacific Ocean where it was used on an experimental basis to broadcast TV programs into Papua New Guinea.

During the four years in which it operated, Hermes confirmed the feasibility of direct-to-home satellite TV broadcasting to small aperture antennas. Hermes also demonstrated that high-power satellites could deliver innovative educational and health services to remote areas. In 1987, the U.S. National Academy of Television Arts and Sciences awarded an EMMY jointly to the CRC and NASA in recognition of the pioneering contribution that Hermes made in advancing the technology of TV broadcasting.

ATS-6 AND THE SITE PROJECT

On May 30, 1974, NASA launched its sixth experimental Advanced Technology Satellite. ATS-6 was the first telecommunication satellite with enough power to broadcast TV signals to small receiving stations. In the U.S., the ATS-6 satellite provided experimental public health and education telecasts to remote rural areas.

In 1975, NASA leased ATS-6 to the government of India for its Satellite Instructional Television Experiment

Fig. 1-8. Author Arthur C. Clarke's home satellite TV receiving system began picking up the SITE TV broadcasts in the summer of 1975.

(SITE), a project destined to transmit TV to remote villages across the subcontinent. In the summer of 1975, Arthur C. Clarke became the first private individual in the world to regularly receive satellite TV broadcasts via a receiving station installed at his home in Colombo, Sri Lanka. The large aperture dish (shown in the accompanying photo) was required to receive the ATS-6 SITE broadcasts to India due to the satellite's low transmit power (by today's standards) and the satellite's use of the UHF frequency spectrum (860 MHz) as opposed to the higher-frequency ranges employed by today's telecommunication satellites.

English amateur radio enthusiast and BBC technician Stephen J. Birkill learned of the Indian SITE project and set out to build his own UHF receiver and antenna for these transmissions. His initial reception in England of the ATS-6 SITE broadcasts took place in December of 1975. From well outside the satellite's footprint, with a five-foot antenna made from screen, he proved that private individuals with limited resources and little outside assistance could bring satellite television into their own homes.

Fig. 1-9. Stephen J. Birkill's experimental Earth station for receiving the SITE broadcasts in the UK. (Courtesy Stephen J. Birkill.)

THE ERA OF THE BACKYARD DISH

Back in the United States, Home Box Office (HBO) and Ted Turner's WTBS began transmitting their respective programming services via satellite to American cable TV systems. Other American program services soon joined HBO and WTBS on Satcom 1, America's first cable TV satellite.

Several people within the American cable TV and aerospace communities began to think about the possibilities of DTH TV reception. In 1976, Stanford Professor H. Taylor Howard designed, built and operated the world's first C-band home satellite TV system and was able to receive the early HBO broadcasts. Bob Cooper, then editor of the U.S. cable TV industry's *CATJ* magazine, obtained an experimental license from the FCC that authorized the installation of a satellite receiving system at his home in rural Oklahoma. In the October 1978 issue of *TV Guide,* he brought the potential of satellite TV to the attention of millions of Americans.

The response to Bob Cooper's 1978 article was nothing short of incredible. Thousands of Americans, many of whom lived in rural areas beyond the reach of terrestrial TV stations, responded with letters asking where they could buy their own satellite TV systems.

This unanticipated demand for satellite TV service gave birth to a whole new industry composed of self-taught experimenters and entrepreneurs rather than engineers and corporate giants. Innovative antenna and receiver designs soon made their appearance, dramatically lowering the cost of those early home satellite TV systems.

THE ROAD TOWARD LEGALIZATION

Many of the major U.S. cable TV programmers, who by then were using satellites in a big way to deliver TV services to cable TV affiliates around the country, accused the early "backyard dish" owners of signal piracy. Their objections were based on the fact that satellite dishes were receiving for free the same TV services for which cable TV subscribers had to pay a fee. When dish owners declared their willingness to pay a subscription fee in exchange for authorized access to satellite programming, premium programmers such as HBO refused to entertain their offers and instead attempted to get the U.S. Congress to pass legislation that would have made home satellite dishes illegal.

The first crack in the door, as far as the viewing rights of C-band dish owners was concerned, occurred in 1982 at the *Satellite TV Technology International* (STTI) convention held in Atlanta, Georgia. The keynote speaker at this event was Robert Wussler, then president of Superstation WTBS. At the conclusion of his keynote speech, I rose from the audience and asked Mr. Wussler whether or not WTBS supported the rights of individuals to receive WTBS by using a satellite TV dish. As far as WTBS was concerned, he responded, home dish owners were welcome to receive Superstation WTBS. This made him the first major TV programming executive to go on record in support of the rights of private individuals to access satellite TV signals.

In 1983, I approached my local congressman to present the case for the legalization of home satellite TV reception. The congressman, who represented a district in Tennessee with a large rural constituency, immediately realized that any legislation which outlawed satellite dishes would be a disservice to millions of Americans with limited access to terrestrial TV channels.

Against the advice of his chief legislative aid, Congressman Albert Gore, Jr. decided to co-sponsor legislation to guarantee the rights of dish owners to access satellite TV program services. In cooperation with the publisher of this book, numerous copies of the 1983 edition of *The World of Satellite TV* were given away to members of Congress and their staffs so that they could make an informed decision concerning what was then a totally new technology. With the help of Senators such as Barry Goldwater, the U.S. Congress passed legislation in November of 1984 that formally legalized home satellite TV reception.

In 1986, MA/COMM introduced new encryption technology that allowed U.S. cable TV services such as HBO and Showtime to control access to their programming and establish a way to bill subscribers so that their satellite services could be marketed directly to American satellite dish owners. Other American programmers soon followed their example, making home satellite TV a hot commodity.

Fig. 1-10. Author Mark Long (left) with SBCA President Chuck Hewitt (center) and former WTBS President Robert Wussler (right).

DBS COMES OF AGE

On the evening of December 17, 1993, I joined more than a hundred other guests at the Guyana Space Center near Kourou, French Guyana to watch Arianespace launch America's first high-power direct broadcast satellite (DBS) into space. Like everyone else in the crowd, I gazed with anticipation across the four kilometers that separated the launch pad from the viewing site. When the moment of ignition finally arrived, the earth trembled and the Ariane 4 rocket roared as a five-pointed star of blazing light hurled toward the heavens. At that moment I sensed that the global satellite TV industry would never be the same again.

In 1994, American companies DIRECTV and USSB inaugurated the first high-powered DBS services for the United States, collectively transmitting more than one hundred and fifty channels of TV programming to American homes equipped with low-cost digital receivers and antennas no more than eighteen inches in diameter. Suddenly home satellite TV was within the reach of millions of American homes that previously only had limited access to TV programming. Welcome to the Twenty-First Century!

THE WORLD OF SATELLITE ENTERTAINMENT

The extent to which you can access the hundreds of satellite TV and CD-quality audio channels already beaming your way will depend on your location and the type of satellite TV system that you elect to purchase. Whenever I mention direct-to-home (DTH) satellite TV systems, I am referring to any receiving system capable of viewing TV services from one or more available satellites. The technical characteristics of the DTH receiving system that you select will depend upon the strength and format of the satellite signals that you intend to receive. The following chapter will provide you with an overview of the many possibilities so you and your family can make an informed choice that best satisfies everyone's viewing needs.

KU-BAND DTH RECEIVING SYSTEMS

The eighteen-inch dish that took the U.S. by storm beginning in 1994 has certainly attracted the attention of TV viewers throughout the Americas. More recently, the cost of a Ku-band "direct broadcast satellite" (DBS) receiving system has plummeted to under the $300 mark. Complete systems are even provided for a low-cost monthly rental fee by one programmer. The price of a professional installation is also relatively low, less than $100 in some instances.

Viewers who are looking for a set program lineup and an unobtrusive dish will be more than satisfied with a fixed-dish installation that can receive a digital program package from a single satellite or constellation of satellites collocated at a single orbital location in the sky. In most cases, the Ku-band digital DTH receiving system consists of an eighteen- or twenty-four inch dish package—called the "outdoor unit"—and an "indoor unit" that the program provider either calls an integrated receiver/decoder (IRD) or a set top box. These receiving systems also come with a plastic access card that plugs into a slot on the front panel of the IRD. Similar in size to a standard credit card, these so-called "smart cards" contain tiny microprocessor chips that provide each IRD with essential security and encryption information without which the IRD cannot function.

Many consumers who resisted the impulse to install a huge dish in their backyard during the 1980s have embraced the much smaller dishes, which receive high-quality digital signals as opposed to the standard "analog" TV signals that have been around for more than fifty years. Small Ku-band dishes can be installed at locations where there simply isn't enough room for a much larger C-band antenna. What's more, it is possible to take your satellite TV system with you for use at alternate locations.

Specialized antenna mounts may be used to install your dish on a recreational vehicle (RV), or even a boat. Portable antenna units also are available that allow the transportation of a Ku-band DTH receiving system from your main residence to an alternate location, such as a vacation home or cabin. Information that appears later in this book will provide all the details you will need to know to realign your dish at alternate locations.

Ku-band DTH Disadvantages. Programming is usually ordered as one or more packages, called "tiers", which offer various basic TV services along with the option to add premium movie networks and/or sports channels for an additional fee. One major downside is that you are limited to viewing the program lineup that the service provider elects to include in its "digital bouquet" of services.

All Ku-band digital DTH dishes are fixed onto a single satellite or group of satellites located at a single orbital location in the sky. Subscribers do not have the option of adding a motor to steer their dish over to view another DBS satellite. The smart card and IRD for each system is unique to a single digital DTH programmer, or in the case of

Fig. 2-1. Components in a small-dish digital DTH receiving system. (Courtesy RCA.)

DIRECTV and USSB, a single program partnership. The receiving system for one platform is therefore incompatible with the equipment used by the other available Ku-band satellites.

Another disadvantage to all Ku-band DTH systems is the service interruptions that will occur at unpredictable times throughout the year. Ku-band signal attenuation may occur whenever there is a high concentration of particles such as rain drops in the atmosphere. Attenuation most often occurs whenever the incoming Ku-band signal must pass through rain-filled layers of the Earth's lower atmosphere. Moderate to heavy downpours will cause "rain fade outages" where the reception of the digital Ku-band TV signal will go from a perfect picture to no picture at all.

The frequency and duration of these outages will vary according to the climate of the area in which you live. This problem will occur more fre-

Fig. 2-2. Large -dish C-band receiving systems have the potential of receiving more channels and thus provide greater freedom of choice. But they also have a greater impact on the home environment.

quently in sections of the country such as Florida that experience a higher incidence of moderate to heavy rainstorms than arid locations such as Arizona. It might sound like a small price to pay for the privilege of using such a tiny dish, but you may have second thoughts if an outage occurs while you are watching a pay-per-view movie or special live event.

LARGE DISH DTH SYSTEMS

One of the early promises of satellite TV when compared to cable was the freedom of choice that it provided the viewer. Cable systems offered potential viewers a set lineup of channels with no guarantee that any one channel would remain a part of the local system's package in the long run. The large steerable dishes that began to make their appearance during the early 1980s gave viewers the option of watching each and every satellite TV service under

the sun. In large measure, this remains one of the key selling points for the installation of a large, steerable dish.

The free spirits in the crowd may be looking for adult entertainment services, access to exotic foreign TV services or the rewards of continuing educational pursuits through one or more of the instructional satellite TV channels linked to major universities and other institutions of higher learning. All of these attractions are pretty much the exclusive domain of today's C-band DTH systems.

Millions of large C-band dishes (typically 2.4m to 3m in diameter) equipped with standard analog IRDs, or even analog-plus-digital IRDs, are currently in use throughout the Americas. Although not as heavily promoted as the available small dish systems, the big dish has much to offer the discerning consumer. Analog-only C-band DTH systems can access more than seventy-five "free-to-air" program services that are transmitted in the clear. Before you get too excited, however, you should know that the available free TV program categories are pretty much limited to religious channels, educational TV services, home shopping venues, broadcast TV network feeds to affiliates and foreign TV stations.

Hobbyists may be attracted to the challenge and unexpected rewards that only a large steerable dish can deliver, such as unrivaled access to dozens of occasional-video channels that offer a cornucopia of unscheduled TV programming. The content of these so-called "wild feeds" includes syndicated TV programs, sports events, and live on-the-scene news reports. Publications such as

Satellite Orbit magazine (http://www.orbitmagazine.com) provide the most current information concerning the availability of wild feed programming, either in their monthly magazine or at their Internet site on the worldwide web.

What's more, almost every major basic and premium program TV service that DBS and cable TV operators offer today also can be received by C-band receiving systems. The available program options include dozens of pay-per-view movie channels and more than 100 free radio stations from across the nation and around the world.

Before you can receive these encrypted or "scrambled" TV signals, however, the C-band DTH system must be equipped with a VideoCipher RS compatible decoder—the signal security platform adopted by the vast majority of basic and premium TV programmers. The viewer must also pay a monthly subscription fee in order to gain access.

C-band program subscription brokers such as Prime Time 24 (http://www.primetime24.com) provide 50-channel packages for under $25 per month as well as the opportunity to order—at an additional cost— extra basic services, premium movie channels and sport networks on an *a la carte* basis. With a C-band dish, you can choose only those channels you wish to view instead of paying for an entire package that may contain channels in which you have little or no interest.

The latest and greatest C-band DTH systems are equipped with a high-tech, multimedia IRD that contains analog and digital satellite tuners and decoders so that your system can receive TV and audio signals in both formats. Manufactured by General Instruments, the "4DTV" IRD (http://www.4dtv.com) contains both the VideoCipher RS and DigiCipher-2 decoders required to unscramble a cornucopia of analog and digital TV signals now carried by the major North American C-band satellites.

With a 4DTV plugged into your C-band dish, an additional 75 free digital C-band TV channels are just a click of the remote away. Moreover, many of the major basic services, premium program channels and pay-per-view movies are also available in the same high-quality digital format that the DBS operators like to pretend is their exclusive domain.

The 4DTV IRD also provides unrivaled access to CD-quality audio services. The major Ku-band digital DTH operators only deliver about 30 CD-quality audio channels per platform and access to these audio services also requires a subscription to at least an entry level program package. The 4DTV can receive more than 75 CD-quality audio services from various C-band satellites without any subscription required.

C-band DTH Disadvantages. Although C-band DTH systems have the potential to offer viewers access to a 600-channel universe, they are not a practical choice for many families. The equipment cost for a typical C-band system is in the $2,000 plus range. The fee for a professional installation can also run up to several hundred dollars. The large (2.4m or larger recommended) C-band antenna required also must be installed at a location that has a clear, unobstructed view of the sky in the direction of the satellite arc.

Although many C-band dish owners think that big is beautiful, the neighbors may not hold the same opinion. In urban and suburban areas, C-band DTH systems are also susceptible to terrestrial interference from local telephone relay stations and other communications installations.

MAKING AN INFORMED CHOICE

Whichever route you elect to take will depend in part on how much room you have for the dish as well as which type of receiving system will deliver the programming categories and brand names in which you are most interested. You may also elect to have both a large steerable C-band dish AND a small Ku-band digital DTH antenna installed at your home so that you can enjoy the best of both worlds. Before you can decide upon the type of satellite system that's best for the entire family, you will need to understand a bit more about how the technology works and the delivery platforms that are available.

UNDERSTANDING THE GLOBAL SATELLITE INFRASTRUCTURE

Today's telecommunication satellites are owned and operated by large organizations and multinational corporations such as INTELSAT, PANAMSAT, General Electric, and Loral Skynet, as well as national telecom companies such as Telesat Canada, Nauhelsat (Argentina) and Embratel (Brazil). These entities assume the initial cost of building, insuring, launching, and operating a satellite, over US $200 million in most cases. They are authorized by government agencies to "sublet" transponder time to other companies on a "common carrier" basis, where the satellite company is the carrier and the user provides the programming.

Broadcasters and cable TV programmers use satellite transponders to instantaneously relay programs to their TV affiliate networks within the region. National broadcast TV networks from countries overseas such as China, France, Germany, Italy, Japan, Portugal, and Spain use satellite transponders to provide regular program feeds to cable TV systems throughout the Americas as well as to transmit DTH services to individual households. The economy of scale afforded by digital video compression is making it possible for countries such as the United Arab Emirates, Croatia, and Thailand, to provide programming feeds into the Americas.

Satellite transponders also relay live and unedited versions of breaking news stories and major sporting events from remote locations back to the main network studios. Many of these occasional-use, wild feeds can be accessed by anyone with a C-band satellite TV system.

Most satellite TV services have been created with both cable and satellite TV viewers in mind. These include news, sports, music videos, educational program services, feature films, pay-per-view (PPV) movies, music concerts, live sporting events, adult TV programming and general entertainment services for the entire family. As the popularity of satellite TV increases in the Americas, additional services for home dish owners inevitably will appear in the years ahead.

Entertainment television isn't the only type of telecommunication traffic that satellites handle. Govern-

Domestic TV Distribution Satellites For North, Central and South America, and the Caribbean

Fig. 2-3. Domestic TV distribution satellites for North, Central &South America, & the Caribbean.

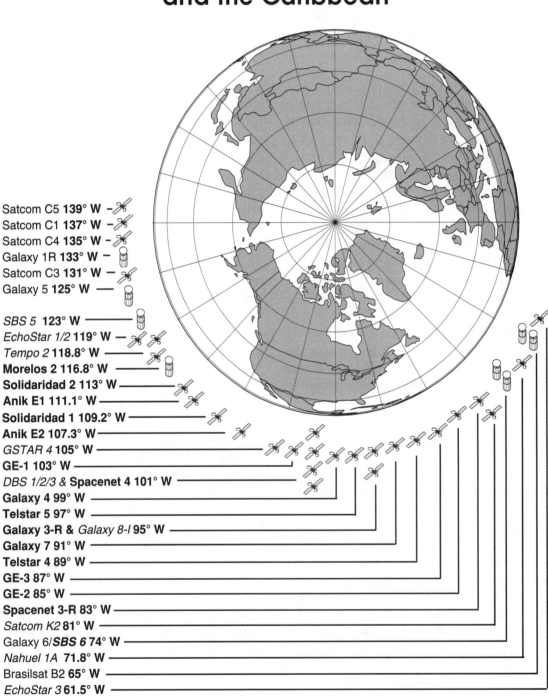

Satcom C5 **139° W**
Satcom C1 **137° W**
Satcom C4 **135° W**
Galaxy 1R **133° W**
Satcom C3 **131° W**
Galaxy 5 **125° W**

SBS 5 **123° W**
EchoStar 1/2 **119° W**
Tempo 2 **118.8° W**
Morelos 2 116.8° W
Solidaridad 2 113° W
Anik E1 111.1° W
Solidaridad 1 109.2° W
Anik E2 107.3° W
GSTAR 4 **105° W**
GE-1 103° W
DBS 1/2/3 & **Spacenet 4 101° W**
Galaxy 4 99° W
Telstar 5 97° W
Galaxy 3-R & *Galaxy 8-I* **95° W**
Galaxy 7 91° W
Telstar 4 89° W
GE-3 87° W
GE-2 85° W
Spacenet 3-R 83° W
Satcom K2 **81° W**
Galaxy 6/*SBS 6* 74° W
Nahuel 1A **71.8° W**
Brasilsat B2 **65° W**
EchoStar 3 **61.5° W**

Satellites in italic type face are Ku-band only;
Satellites in plain type face are C-band only;
Satellites in bold type face are dual C/Ku-band.

(Not shown: inclined orbit satellites Telstar 303, Anik C1, Satcom K2, Brazilsat A1, and SBS 4 at 120, 118.6, 81, 79 and 77° West, respectively.

ment agencies, banks, and major corporations use satellites to relay computer data at speeds of several million bits of information per second, send mail electronically, and carry on private telephone conversations. Large merchandisers such as Walmart and 7-Eleven use satellites to link their corporate headquarters with hundreds or even thousands of retail outlets. Services include in-house videoconferencing, credit card transaction processing, inventory control and other data communications services. What's more, the Internet has become one of the largest users of satellite capacity both here in North America and abroad.

A PROGRAMMING OVERVIEW

Almost all of the programming categories described below are available to owners of either digital or analog satellite TV receiving systems. For those cases where the program category is available to only one type of system, we will alert you to the fact by placing a digital (left) or analog (right) TV icon at the beginning of that entry.

Feature Films and Pay-Per-View Movies. Numerous movie channels are available that deliver popular and critically acclaimed films in a tantalizing array of languages. Some of these channels, as well as numerous general entertainment services,

Fig. 2-4 DTH Satellite System Comparison Chart.

DTH SATELLITE SYSTEM COMPARISON CHART (As of April, 1998)	C-BAND DTH/ ANALOG IRD	C-BAND DTH/ 4DTV ANALOG & DIGITAL IRD	DIRECTV & USSB	ECHOSTAR DISH NETWORK	PRIMESTAR DTH (GE-2)
SYSTEM PARAMETERS					
FREQUENCY BAND:	C-band	C-band	Ku-band	Ku-band	Ku-band
DISH SIZE (may vary with location):	1.8~3.0m	1.8~3.0m	45cm	45cm	65~100cm
IRD TV STANDARD:	analog	both	digital	digital	digital
EQUIPMENT COST:	$2,000	$2,500	>$300	>$300	NONE(*)
INSTALLATION COST:	$500	$500	$100	200	$150
RAIN FADE OUTAGES:	NO	NO	YES	YES	YES
POTENTIAL TI PROBLEMS :	YES	YES	NO	NO	NO
TOTAL NO. OF TV/AUDIO CHANNELS AVAILABLE:	370+	600+	200+	167+	166+
FULL TIME FREE TV:	90+	160+			
WILD-FEED (o/v) FREE TV:	50+	50+	NONE	NONE	NONE
FREE AUDIO SERVICES					
MONO & STEREO FM:	100+	100+	NONE	NONE	NONE
CD-QUALITY:		70+	see below	see below	see below
SUBSCRIPTION CHANNELS					
C-BAND ANALOG:	130+	130+			
C-BAND DIGITAL:		90+			
KU-BAND DIGITAL:			200+	167+	166+
PROGRAM CATEGORIES:					
PREMIUM MOVIES:	20+	40+	30		
PAY PER VIEW MOVIES:	6	36	55	10	15
SPORTS NETWORKS:					
CD QUALITY AUDIO:			31	31	30
ADULT TV CHANNELS:	7	10	2		
* Equipment rental, service and maintenance is part of PRIMESTAR's monthly subscription fee.					

also offer drama, mini-series, comedy, talk shows, music videos, documentaries and other entertainment fare.

Impulse pay-per-view movies (IPPV) and special events are also available. PPV movies are provided well in advance of when they will air on premium cable channels such as HBO and Showtime, or on any of the broadcast television networks. PPV movie services are currently available to owners of both digital and analog TV systems. To receive PPV services, the DTH system must be equipped with hardware accessories such as a decoder and a modem that connects to the home's telephone line. These accessories may either be built into the system's IRD or come as accessories that plug into the DTH system's receiver.

The viewer can select a PPV program from an on-screen menu and the modem automatically dials the programmer's toll-free number and provides all the information required by the programmer's billing center for the event that you have selected. The programmer then sends a coded message to your receiving system to instruct the system's decoder to switch on for the duration of the event that you have ordered.

Sports. With several full-time national sports channels, regional sports networks and national broadcast networks with regular sporting event coverage, it is now possible to view a dazzling array of athletic competition from around the world, including live coverage of soccer, basketball, football, hockey, tennis, volleyball, and baseball games. There are literally hundreds of hours of sports coverage every week relayed via satellite, more than enough for

even the most dedicated fan. Dish owners may also subscribe to packaged sports services that allow the viewer to customize his or her subscription to follow their favorite sports teams.

"Wild feeds" that can relay syndicated TV programs or the broadcast TV transmissions from distant sports arenas or sites of breaking news events to their respective home TV studios, also are available to owners of large-dish analog TV systems who know when and where to look.

Satellite TV viewers who are located outside of the coverage area or "footprint" of the North American satellite or satellites carrying major sports services may still have potential access to sports events of international importance. There is a reasonable chance that any major international sports event will be available to satellite dish owners via one of the international INTELSAT or PANAMSAT satellites, which are commonly used to relay the Olympics, the Pan American Games and World Cup Soccer matches to an international audience.

News and Information Channels. BBC World Service Television, CNBC, CNN and CNN International are just a few of the news channels that are available from satellites serving the Americas. The various national broadcast networks using satellites within the region also transmit regular daily news programs.

Some of the more interesting satellite TV news transmissions do not occur on a scheduled basis. Unedited and unexpurgated, these raw video feeds let you watch everything, not just what the networks want you to see.

With a little extra effort and some patience, you can explore these occasional video (o/v) channels and view live news feeds that place you right in the middle of the action. Moreover, international news services such as Reuters TV and WTN offer daily news feeds to the region covering major international and regional events.

Transportable satellite uplink stations are also used by global and national news organizations to transmit live on-the-scene reports back to their main TV studios. Gaining direct access to these satellite news feeds puts you on the cutting edge of worldwide developments. The region's many domestic satellites, as well as the international INTELSAT and PANAMSAT satellites often serve as effective distribution hubs for relaying overseas news feeds. Broadcasters such as ABC, CBS, NBC, CBC, NHK, and the BBC use these satellites either to send program feeds to their respective home studios or to relay them to affiliated broadcast stations within the region.

Family Entertainment Networks. There are numerous family entertainment networks which provide a wide range of programming options. These include live sports events, reruns of sitcoms, soap operas, dramas, and other popular network TV programs now in syndication, cartoons for the children, music videos, nature documentaries, computer tips, game shows, and movies that are no longer in heavy rotation on the premium movie channels described above.

Music Video Channels. MTV (Music Television) started it all. Now there are a wide variety of music video services covering different demographics and music genres.

Superstations. During the 1980s, independent TV stations such as WGN (Chicago) and WTBS (Atlanta) achieved national fame by uplinking their services onto satellites so that cable TV systems throughout the country could carry their program services. These satellite broadcast channels, called "Superstations" by the industry, commonly acquire the national rights to TV programs so that they can legally offer them to viewers throughout the country.

Soft-core Adult Entertainment Networks. A few of the digital DTH service providers provide soft-core adult entertainment services such as the Playboy Channel and Spice. These services are optional if you do not want to have adult entertainment programming available in your household.

Hard-core adult movie channels. The steamier adult entertainment fare is only available to large-dish analog TV systems equipped with decoders that are compatible to the encryption systems used to prevent youngsters from viewing these programs.

Religious Program Services. C-band satellites also deliver a variety of religious program services. These include Eternal Word TV Network (EWTN), Sky Angel and Trinity Broadcasting Network (TBN).

Foreign Language TV Services. The major digital DTH service providers offer viewers the option of selecting a foreign language sound track for a few of the available digital program services. In the U.S., alternate Spanish-language audio channels can be selected by the viewer for popular North American movie services such as Cinemax, Encore, HBO, Showtime, and STARZ. Foreign-language TV services such as ART (Arabic), Antenna (Greek), RIA (Italian), NHK and TV Japan (Japanese) and MTV Latin, Univision and Telemundo (Spanish), are also available on tap from select digital DTH providers. Canadian digital DTH viewers also benefit from access to both English- and French-language versions of their favorite TV program services.

Be forewarned, however, that the foreign-language program fare offered by the major digital DTH systems represents just the tip of the iceberg. The program networks from more than two dozen additional countries worldwide are now available via satellites covering the Americas. These include national TV networks from China, France, Germany, Mexico, Portugal, Spain, Thailand and the United Arab Emirates.

Audio Programming. Many satellites also deliver numerous audio services with programming that ranges from classical and current pop hits to easy listening and heavy metal rock. News and talk radio services also are available. Many of these audio services are either offered in a high-fidelity stereo format or in a CD-quality digital stereo format.

Internet and Other Data Services. Satellite dishes aren't just for receiving TV and audio programming these days. Satellite-based Internet service providers such as DIRECPC (http://www.direcpc.com) offer lightening quick data downloads from the worldwide web (WWW). DIRECPC and DIRECTV also have teamed up to provide both digital DTH and Internet access from a single satellite dish. Called the DirecDuo, this system actually receives two satellite platforms at once, with one platform providing the Internet downloads and the other delivering digital DTH.

News and financial information services from X*PRESS and Reuters also are available. To receive any of the above-mentioned services, the dish must be pointed at the satellite carrying the service and the cable from the dish must connect to either a PC card that plugs into an IBM PC compatible computer system or to a special stand-alone receiver/decoder.

SATELLITE PROGRAM GUIDES

With such an awesome array of programming available, a program guide becomes an essential tool to keep you from endlessly searching for the program that you want to view. Similar to the TV listings in your local newspaper, satellite guides attempt to list the available programming on an hour-by-hour basis. They also provide readers with up-to-date "satellite program grids" that show you the programming logs for each of the available services. Check out your local newsstand to determine the type of satellite TV programming guide which best fulfills your individual needs. If you subscribe to a program package, whether analog or digital, the program provider may also supply a monthly guide to its subscribers.

Video and audio services periodically change their transponder assignments on a given satellite or move from one satellite to another. Information that is current one day may be totally obsolete the next. It therefore is not my intention herein to

Fig. 2-5. DirecTV's DBS-1 satellite prior to launch. (Courtesy Hughes.)

customized lists of "favorite channels."

US DBS PLATFORMS

DirecTV (DBS-1, DBS-2 & DBS-3), EchoStar (Echostar 1 & 2), and USSB (DBS-1) currently operate high-power broadcast satellite service (BSS) systems serving subscribers in the United States. PrimeStar plans to begin offering a new high-power broadcast satellite service of its own in addition to a medium-power Ku-band service already underway. All of these BSS systems conform to plans adopted by the World Administrative Radio Conference (WARC) held in 1977 and the Regional Administrative Radio Conference (RARC) for the Americas held in 1983. Both conferences, which took place under the auspices of the UN's International Telecommunication Union (ITU), envisioned that only high-power satellites (up to 260 watts per transponder) operating with a minimum of nine degrees of separation between satellite systems serving the same general area could be used effectively to transmit TV services directly to individual homes.

The RARC-83 satellite broadcasting plan for the Americas assigned specific transponder frequencies (12.2 to 12.7 GHz), polarization formats (right-hand or left-hand circular) and orbital locations to each country within the Americas. These specifications were designed to prevent interference between the BSS systems of adjacent countries. To date, only the United States has implemented RARC-compliant sys-

offer a satellite-by-satellite list of all the available programming services. This is a task more suitable to Internet web sites and magazines, which can update their lists on a daily, weekly or monthly basis.

To make channel surfing easier, the new "digital bouquet" programmers electronically deliver program guides over their satellite systems that break down their various offerings into individual categories such as movies, sports and special interest. An on-screen electronic program guide (EPG) provides all the details concerning what's coming up next.

Users can scan the EPG and select program choices by using the IRD's hand-held remote control. In some cases, you can even superimpose the EPG onto the TV screen so that you can continue to watch a program while you check out the guide. Subscribers can also use the remote control to instantly order pay-per-view movies, set parental controls over program content and spending limits, as well as create one or more

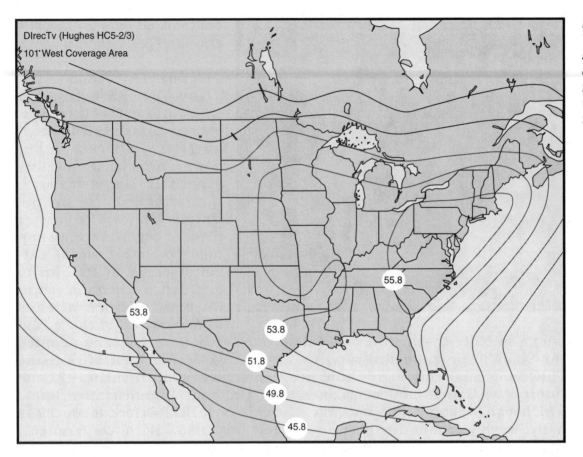

DIrecTv (Hughes HC5-2/3)
101˚West Coverage Area

53.8
53.8
55.8
51.8
49.8
45.8

*Fig. 2-6.
Typical DBS
satellite
coverage
from 101°
west longi-
tude.*

tems in the Americas. Beginning in 1999, however, Canada will have its own high-power DBS satellite in orbit at 91 degrees west longitude.*

(*Although the International Telecommunication Union (ITU) has adopted the term "broadcast satellite service" (BSS) as part of its nomenclature to describe high-power satellite TV broadcasting in compliance with specifications developed under international agreements, the term "direct broadcast satellite" (DBS) has achieved a much higher level of popularity with the general public. Therefore, whenever I use the term DBS in this book I am referring to satellite systems and/or services that fully conform to the "broadcast satellite service" (BSS) specification set forth in the ITU regulations. Whenever I use the term DTH I am referring to any satellite TV service, whether digital or analog, which can be directly received from any communications satellite. Those satellite purists who may enjoy quibbling over such matters, please take note!)

DIRECTV

On December 17, 1993, Arianespace launched the first high-power direct broadcast satellite (DBS) for the United States from the European Space Agency's launch facilities in French Guyana, South America. It was my distinct privilege to be on hand that evening to watch DirecTV's DBS-1 satellite blast off to its new home in the sky. Since the inaugural launch of DBS-1, DirecTV has deployed two additional satellites known as DBS-2 and DBS-3 on August 2, 1994 and June 9, 1995, respectively. All three satellites are collocated in the immediate vicinity of the orbital location of 101 (100.8, 101 & 101.2) degrees west longitude.

DirecTV (Internet web site at: http://www.directv.com) currently

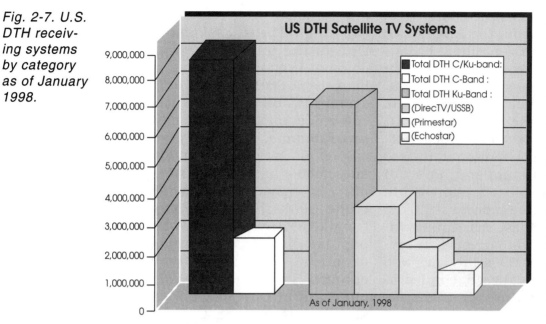

Fig. 2-7. U.S. DTH receiving systems by category as of January 1998.

US DTH Satellite TV Systems

Legend:
- ■ Total DTH C/Ku-band:
- □ Total DTH C-Band :
- ▨ Total DTH Ku-Band :
- ▨ (DirecTV/USSB)
- ▨ (Primestar)
- □ (Echostar)

As of January, 1998

delivers more than 175 channels of digital TV programming to homes and businesses equipped with "digital satellite system" (DSS) home receiving units. The same DSS system can also access twenty-six more channels from USSB, a second digital DTH provider that owns five transponders on DirecTV's DBS-1 satellite.

The DSS system consists of an 18-inch dish, a digital set-top decoder box and hand-held remote control. Compatible DSS hardware is currently available under the following brand names: GE, Hitachi, Hughes Network Systems (HNS), Magnavox, Memorex, Panasonic, ProScan, RCA, Radio Shack OPTIMUS, Sony, Toshiba, and Uniden.

DirecTV's DBS satellite uplink and digital broadcast facilities are located at the Castle Rock Broadcast Center (CRBC) in Castle Rock, Colorado. All DirecTV transmissions originate from this location. Much of the programming that the CRBC transmits or "uplinks" to the DBS satellite system comes from other

North American satellites or is delivered via fiber-optic links. Other programming resources are accessed from digital video-tape by means of an automatic tape server system. The CRBC digitally compresses and encrypts all of these program resources before transmitting the signals to the DBS satellite constellation.

DirecTV programming includes popular cable networks such as Disney Channel, CNN, ESPN and TNT, as well as up to fifty-five pay-per-view movie choices from the major Hollywood studios each and every day. You can use the indoor unit's remote control to order any one of these PPV movies, which appear as often as every 30 minutes. It is also possible to preview the first few minutes of any PPV movie for free.

Professional and collegiate sports services and events are also available, including NFL SUNDAY TICKET, NBA LEAGUE PASS, NHL CENTER ICE, and Major League Baseball EXTRA INNINGS. Also available: Total Choice "Gold" and "Platinum" packages that deliver sporting event packages from around the nation. Soft-core adult program services such as PLAYBOY TV and SPICE are also available on an *a la carte* basis.

DirecTV also delivers Eastern affiliate stations from the ABC, CBS, FOX, NBC and PBS national broadcast networks, as well as Western affiliate stations from ABC, CBS, and NBC. These channels are made available to those viewers who either are unable to receive local networks with a conventional terrestrial rooftop antenna and have not subscribed to cable television within the previous three months.

DirecTV provides scheduling information to its subscribers in several ways. DIRECTV's Internet web site (http://www.directv.com) provides monthly listings for all movies presented in a panoramic "letterbox" format as well as the pay-per-view movies to be offered. The DSS receiving system's on-screen electronic program guide also lists all PPV movies and events 72 hours in advance of their time slots. At the beginning of each month, DirecTV subscribers also receive a listing of all the forthcoming movies and special events.

US SATELLITE BROADCASTING (USSB)

Founded by Twin Cities-based Hubbard Broadcasting, Inc., USSB (http://www.ussb.com) filed for a Direct Broadcast Satellite (DBS) license with the Federal Communications Commission back in 1982 when few broadcast industry experts believed that the satellite delivery of programming was a viable venture. By the late 1980s, however, the development of new digital compression technologies had dramatically increased the economics of DTH program delivery.

In 1991, USSB and Hughes Electronics Corporation agreed to jointly develop a common distribution system for DBS service and to use the same digital broadcast technologies. Due to the joint development of a common digital DTH delivery system between USSB and DirecTV, DSS system owners now have access to a combined total of 210 channels of programming.

Each month more than 900 different movies air on USSB's eighteen different movie channels. These include "multichannel" programming from HBO (five channel) featuring popular films from Hollywood, two channels of HBO Family, a four-channel Showtime multiplex, three channels of Cinemax and two versions of The Movie Channel. Also available: FLIX provides favorite movies of the 60s, 70s, and 80s and Robert Redford's Sundance Channel premiers the best in independent film. Both of these services offer movies uncut and commercial free.

ECHOSTAR

In 1987, EchoStar Communications Corporation filed for a Direct Broadcast Satellite (DBS) license with the FCC and established the EchoStar Satellite Corporation (http://www.echostar.com) to build, launch and operate a new series of DBS satellites. The FCC subsequently awarded EchoStar with a DBS orbital slots at 119, 61.5 and 148 degrees west longitude.

On December 28, 1995, a Chinese Long March 2E rocket successfully launched the EchoStar I satellite to orbit. On March 4, 1996, EchoStar's "DISH Network" began broadcasting digital DTH services to customers throughout the United States.

Arianespace successfully launched the company's second DBS satellite, EchoStar II, on September 10, 1996.

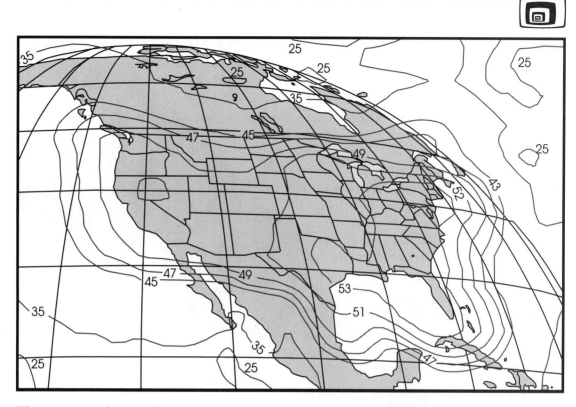

Fig. 2-8. Typical EchoStar satellite coverage from 119° west longitude.

This second satellite is currently collocated with EchoStar I at 119 degrees west longitude. EchoStar III was launched from Cape Canaveral, Florida on October 5, 1997 and is now located at 61.5 degrees west longitude. The three satellites have a combined capacity of 48 high-power transponders that collectively can deliver more than 200 channels of digital video, audio and data services to homes throughout the continental United States.

A Russian Proton rocket is scheduled to launch EchoStar IV from the Baikonur Cosmodrome in the Republic of Kazakhstan in 1998. From a new orbital position at 148 degrees west longitude off the West Coast of the United States, the satellite will be able to provide the DISH Network additional capabilities, including high-power coverage of Alaska and the Hawaiian Islands.

The DISH Network offers a lineup of basic and premium program services that is similar to what DIRECTV and USSB offer from their joint satellite platform at 101 degrees west longitude. What sets the DISH Network apart from its competitors, however, is its commitment to carrying a wider spectrum of foreign TV networks and ethnic program fare.

One of the major reasons why cable TV remains popular in many urban areas is that it offers viewers access to all the local channels along with multitiered packages of basic and premium TV services. The DISH Network intends to counteract this drawback to digital DTH by providing feeds of local TV stations into the top TV markets on the east coast using its EchoStar III satellite at 61.5 degrees west longitude. The only drawback to this plan is that viewers on the east coast will need two dishes to receive both the local channels and the DISH Network's package of basic and premium program services.

PRIMESTAR DTH

Launched in 1994 as America's first digital satellite TV service provider, PrimeStar is currently the only digital DTH service provider in the U.S. that doesn't require the purchase of equipment. An equipment rental fee, which includes all service and maintenance, is included in the monthly subscription fee charged to each PrimeStar subscriber.

PrimeStar's digital DTH service currently offers a programming lineup consisting of 160 channels, most of which are also offered by its previously mentioned competitors. The GE-2 satellite located at 85 degrees west longitude currently broadcasts all of these services.

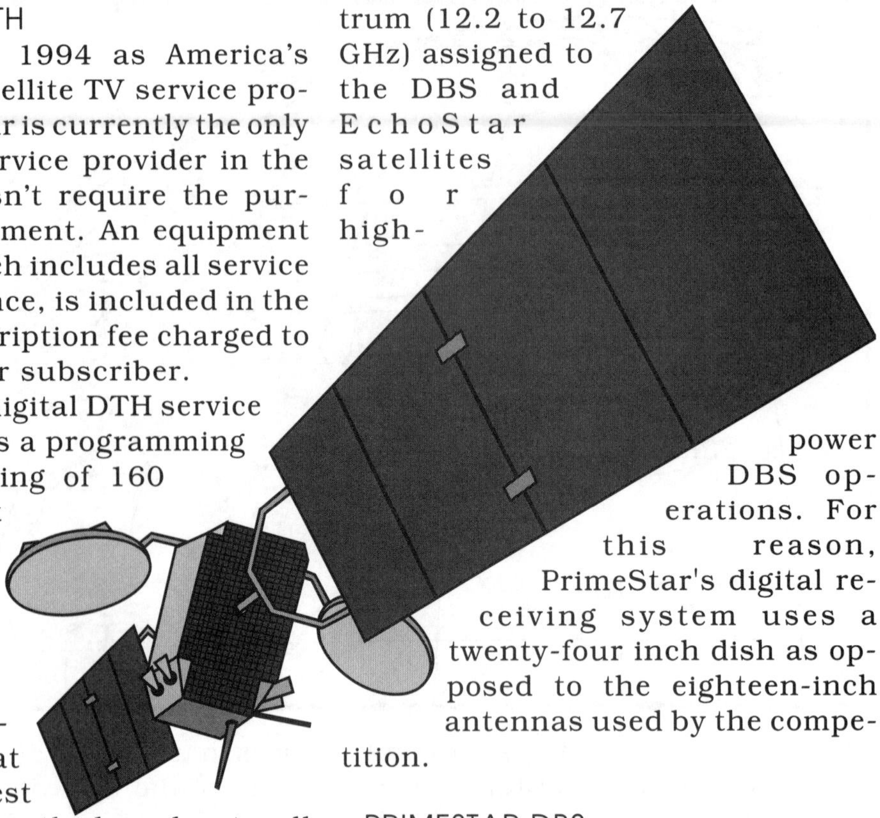

GE-2 is a medium-power spacecraft which operates in a portion of the Ku-band spectrum (11.7 to 12.2 GHz) that is adjacent to the spectrum (12.2 to 12.7 GHz) assigned to the DBS and EchoStar satellites for high-power DBS operations. For this reason, PrimeStar's digital receiving system uses a twenty-four inch dish as opposed to the eighteen-inch antennas used by the competition.

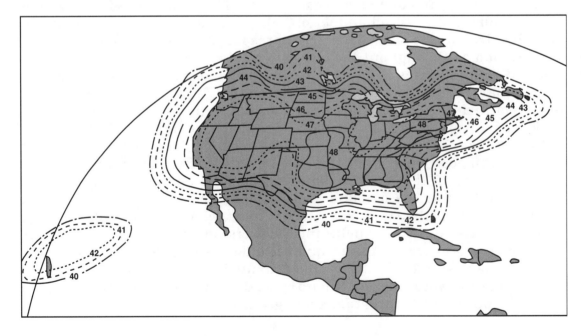

Fig. 2-9. Line drawing of the GE-2 spacecraft.

PRIMESTAR DBS

While continuing to fully market and support its current medium-power service on GE-2, PrimeStar (http://www.primestar.com) also is moving forward with its plans to

Fig. 2-10. GE-2 Ku-band coverage beam from 85° west longitude.

launch high-power DBS services at 119 and 110 degrees west longitude. Eleven high-power DBS transponders on the Tempo-1 satellite at 119 degrees west longitude are currently available for a new 120-channel service that can be received by an eighteen-inch dish.

Manufactured by General Instruments, the new digital IRDs for this service will feature an interactive program guide allowing viewers to watch a program and peruse the guide simultaneously, configure the guide by programming type and favorite channels, and set timers without the need to access a separate menu. The new IRD will also contain an analog satellite TV tuner that will offer video service providers the opportunity to transparently upgrade subscribers from analog to digital programming at some point in the future. Also available: a personal message mailbox and the provision of data and interactive services.

The new General Instrument receiver will also be fully compatible with a second high-power DBS service that PrimeStar intends to launch at 110 degrees west longitude, contingent upon U.S. government approval of a recent agreement reached between PrimeStar and MCI/News Corporation. Under the agreement, MCI and News Corporation would transfer to PrimeStar its DBS frequencies at 110 degrees west longitude together with two high-power DBS satellites currently under construction.

CANADA'S DBS SATELLITE

Telesat Canada (http://www.telesat.ca) is currently supervising the construction of a new high-power DBS satellite to be located at 91 degrees west longitude. The new spacecraft will carry thirty-two transponders operating in the 12.2 to 12.7 GHz frequency band in full compliance with the ITU's DBS regulations for the Americas.

Two Canadian digital DTH operators, ExpressVu and Electronic Digital Delivery (EDD), have announced their intentions to migrate to the new Canadian DBS platform. ExpressVu currently offers Ku-band digital DTH services from Telesat Canada's Anik E2 satellite at 107.3 degrees west longitude. EDD is a new service that intends to use Canada's new DBS platform to deliver an electronic "video store in the sky" that will allow viewers to order and download movies and other types of programming. The company estimates that the download time for a 100-minute movie from the Canadian DBS satellite will be about five minutes.

DIGITAL DTH IN CANADA

The operators currently offering DTH services in Canada include: ExpressVu, HomeStar, Star Choice, AlphaStar Canada and Power DirecTV. U.S. DBS operators are prohibited from marketing DTH services in Canada under Canadian governmental regulations.

EXPRESSVU. ExpressVu Inc. (http://www.expressvu.com) was the first company in Canada to receive a license to provide a full range of DTH services to Canadians in every region of the country. The company holds the Canadian rights to the same digital DTH technology that EchoStar uses in the USA as well as the exclusive use of the DISH Network brand name in Canada.

ExpressVu currently uses fourteen

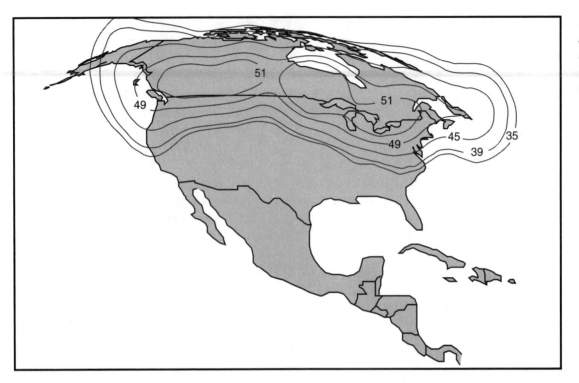

Fig. 2-11. Anik E2 Ku-band national beam coverage from 107.3° west longitude.

Ku-band (11.7 to 12.2 GHz) transponders on the Anik E-2 satellite located at 107.3 degrees west longitude to transmit as many as 100 digital video and audio channels to subscribers throughout Canada. With the transition to the new Canadian DBS satellite in 1999, ExpressVu will have the ability to transmit as many as 180 TV channels through seventeen transponders on the new DBS platform.

ExpressVu programming is transmitted to the Anik E2 satellite from ExpressVu's digital broadcast center in North York, Ontario. The satellite then transmits the signal directly to subscribers equipped with 24-inch (60-cm) dishes, a DISH Network set-top box and a smart card that has been purchased from an authorized ExpressVu dealer.

STAR CHOICE. Digital DTH service provider Star Choice (http://www.starchoice.com) also currently uses Ku-band capacity on the Anik E2 satellite at 107.3 degrees west longitude to serve its customer base. Star Choice offers customized program packages that allow each subscriber to choose a lineup that can be tailored to fit individual interests. Star Choice currently offers 126 video and audio channels of which 107 channels are available to subscribers in the eastern provinces, including a diverse offering of French, and 95 channels for subscribers in the western provinces.

DIGITAL DTH FOR THE AMERICAS

In Central America, the Caribbean and South America, there are several program providers that are using digital DTH technology to relay video and audio programming to subscribers either on a regional or domestic basis. The following section provides an overview of the major players and systems. None of the services listed below have adopted system designs that conform to the

Fig. 2-12. Existing and planned Ku-band digital DBS and DTH satellites for the United States and Canada. (PrimeStar's use of the News Corp./ MCI orbital slot and frequencies at 110° west longitude is subject to U.S. government approval.)

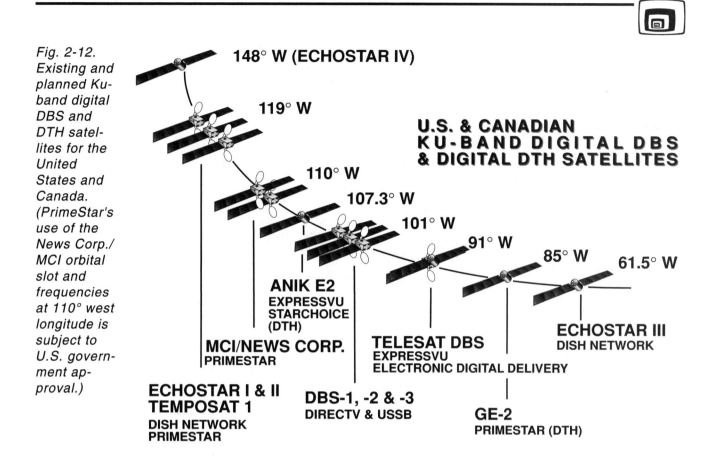

148° W (ECHOSTAR IV)

119° W

U.S. & CANADIAN KU-BAND DIGITAL DBS & DIGITAL DTH SATELLITES

110° W

107.3° W

101° W

91° W

85° W

61.5° W

ANIK E2
EXPRESSVU
STARCHOICE
(DTH)

ECHOSTAR III
DISH NETWORK

MCI/NEWS CORP.
PRIMESTAR

TELESAT DBS
EXPRESSVU
ELECTRONIC DIGITAL DELIVERY

ECHOSTAR I & II
TEMPOSAT 1
DISH NETWORK
PRIMESTAR

DBS-1, -2 & -3
DIRECTV & USSB

GE-2
PRIMESTAR (DTH)

ITU regulations for broadcast satellite service (BSS) operations.

GALAXY LATIN AMERICA

In 1995, Galaxy Latin America (GLA) was created to bring DirecTV's digital DTH service to the Caribbean and Latin America (http://www.directvnet.com). GLA is owned by Hughes Electronics; Venezuela's Cisneros Group of Companies; Brazil's Televisao Abrial; and Mexico's MVS Multivision. Through local alliances in each Latin American country, GLA launched its DirecTV service in 1996. Today, the service is available in twelve countries, which represents 75 percent of the potential market. Before the end of 1998, the company expects to reach more than 95 percent of the market after the launch of service in Argentina and the rest of the countries in the region.

In December of 1995, Galaxy III-R satellite was launched to 95 degrees west longitude. This

Fig. 2-13. Galaxy VIII-i Ku-band coverage beams from 95° west longitude.

dual-band satellite was designed so that its C-band capacity could be used to serve the North American market, while at the same time its entire Ku-band capacity could be employed to provide interim digital DTH services in Latin America and the Caribbean. Twelve Ku-band transponders currently deliver Portuguese-language programming to Brazil, while the remaining twelve transponders relay Spanish-language programming to the rest of Latin America.

GLA's DirecTV service currently offers a total of 164 video channels and 66 audio channels in Spanish and Portuguese. Subscribers can receive signals from the Galaxy satellite located at 95 degrees west longi-

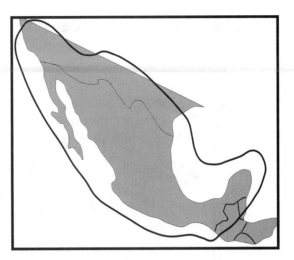

Fig. 2-14. Solidaridad Ku-band coverage beam for Mexico.

tude using antennas ranging from 60cm to 1.1m in diameter depending on location. Services are broadcast from four program centers located in Long Beach, California; Mexico City, Mexico; Caracas, Venezuela; and Sao

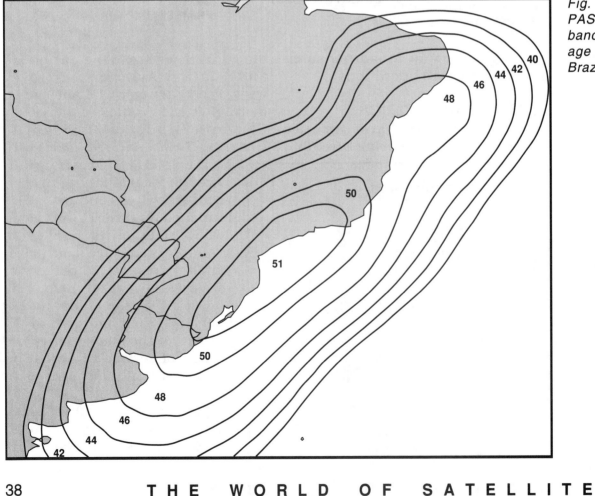

Fig. 2-15. PAS-3R Ku-band coverage beam for Brazil.

Paulo, Brazil. A fifth broadcast center is being built in Buenos Aires, Argentina.

On December 8, 1997, Galaxy VIII-i was launched and collocated with Galaxy III-R at 95 degrees west longitude. Galaxy VIII-i, which has the capacity to offer 300 digital TV channels to viewers throughout the region, is a high-power (118 watts) satellite with thirty-two Ku-band transponders. The new satellite is slated to become GLA's dedicated platform for the region.

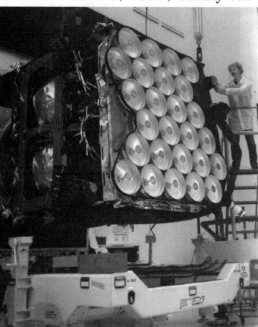

Fig. 2-16. Close-up view of the Solidaridad 2 satellite prior to launch. In addition to carrying C- and Ku-band payloads, Solidaridad 2 also has a 26-element array for mobile communications.

SKY LATIN AMERICA

The other digital DTH operator for the region is Sky Latin America, a partnership between Mexico's Grupo Televisa, Rupert Murdoch's News Corporation, Tele-Communications International, Inc. (TCI), and Brazil's TV Globo. Sky Latin America currently uses capacity on two satellite platforms to reach subscribers within the region. The Solidaridad 2 satellite located at 113 degrees west longitude provides service within Mexico, while the PAS-3R and PAS-6 satellites collocated at 43 degrees west longitude deliver digital DTH services into South America. At the time of writing, Sky Latin America was primarily serving subscribers in Mexico and Brazil.

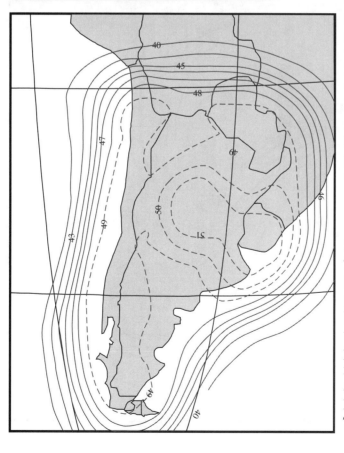

Fig. 2-17. Nahuel Ku-band national beam coverage from 71.8° west longitude.

TV DIRECTA AL HOGAR

In Argentina, a new digital DTH service provider called Television Directa al Hogar (TDH) is using spot beam capacity on the Nahuel-1 Ku-band satellite at 71.8 degrees west longitude. The primary purpose of the service is to reach subscribers in rural areas of the country who have limited or no access to cable TV programming.

THE HISPASAT SATELLITES

Spain's multi-purpose Hispasat (http://www.hispasat.com) satellite system also provides a DTH service for the Americas, with both digital and analog services available to antennas ranging from 90 to 120cm in diameter. The Hispasat 1 and 2 satellites, which are collocated at 30 degrees west longitude, generate a broad coverage beam for the Americas, reaching all the way from Canada to Argentina. Hispasat also will be launching a third satellite in 1999 that will provide additional Ku-band capacity for services into the Americas.

The Hispasat digital DTH bouquet includes four TV services produced by Radiotelevision Española. Hispasat also offers an analog DTH service from Television Española that includes auxiliary radio services from Spain's Radio Nacional and Radio Exterior. Further information on how to receive signals from the Hispasat satellite system from locations in the Americas appears in the Hispasat section of Chapter 11.

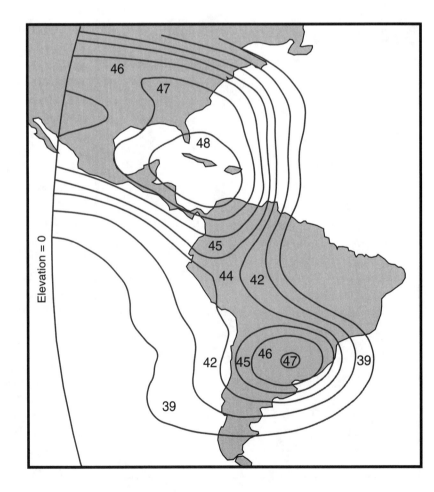

Fig. 2-18. Hispasat Ku-band Americas beam coverage from 30° west longitude.

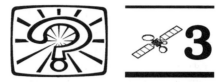

UNDERSTANDING YOUR SATELLITE TV SYSTEM

Now that we have taken a look at the satellites and all the TV services that are up there, you should be ready for a ballpark tour of the technical basics. You do not have to understand all the technical details contained in this chapter to enjoy your system. It's similar to operating your regular TV set—you don't have to know anything about the antenna or cable. It can be as simple as pushing a button corresponding to your favorite channel or activating the electronic program guide (EPG) built into your set-top box. But you may appreciate an overview of the internal workings of your system. First of all, let's take a look at a block diagram of how all the individual components in the system fit together.

THE ANTENNA

The microwave signals transmitted by the satellites are very weak in strength when they finally arrive at your home.

Fig. 3-1. Digital DTH satellite system block diagram.

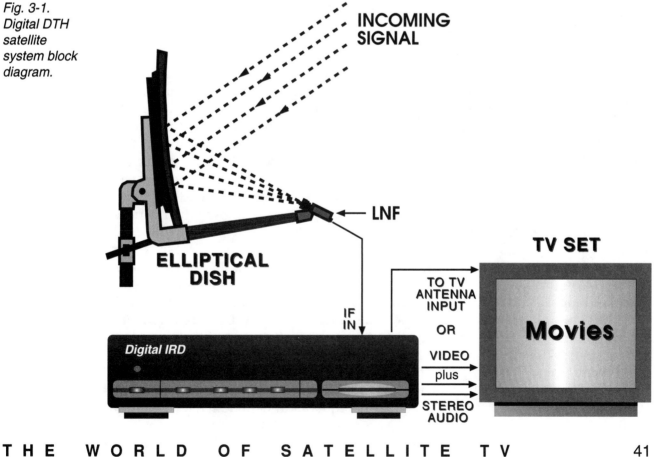

The antenna is typically less than 3.28 feet (1 meter) in diameter for Ku-band DTH receiving systems and between 6 and 10 feet (1.8 to 3 meters) in diameter for C-band DTH receiving systems. The dish captures as much signal as possible and reflects it to a common point towards the front and center of the dish, called the focal point. The dish antenna must be accurately pointed at the satellite in order to gather enough signal to make a picture appear on your TV screen. Being off by more than an inch or two can make the difference between a good picture and no picture at all. A metal device called the feedhorn is located at the focal point and helps to gather up the reflected microwave signals, conducting them back to the first stage of electronics with a minimal amount of signal loss.

THE LNB

The first stage of electronics is called the low noise block downconverter or LNB. It looks like a metal box with a rectangular opening in the front of it. Centered within this opening is a short metal probe less than an inch in length. This probe is the actual receiving antenna. It picks up the signal that is reflected off the dish and conducts it to the first electronic circuit of the LNB.

Most Ku-band digital DTH systems use a combination of a feedhorn and LNB that is called an LNF, short for low noise feed. Use of this unit eliminates an outdoor connection whereby moisture can seep into the system and create reception problems.

The LNB consists of two distinct stages. The LNB's initial stage is comprised of several microwave amplifier circuits, one right after the other. They multiply the incoming microwave signal of less than a millionth of a volt by a factor of 100,000 or more.

This amplified satellite signal is then sent on to the second stage of the LNB called the downconverter. The downconverter is a specialized electronic circuit that reduces the incoming microwave signal to an intermediate frequency range called the "IF".

A microwave mixing circuit inside the downconverter combines the incoming satellite signal with a second signal generated by the LNB's internal local oscillator (LO), also called a dielectric resonant oscillator (DRO). Dielectric materials such as Teflon or ceramics are used since their physical characteristics cause the circuit in which they are present to resonate in the vicinity of a predetermined frequency. The DRO is also pretty stable; this keeps your LNB from drifting off frequency to look for UFOs, black holes and other freaky space stuff.

The DRO circuit is called an oscillator because it electronically vibrates back and forth to produce a microwave signal. When the incoming and DRO signals beat or "heterodyne" together, a new block of lower frequencies is created. This "intermediate frequency" band contains all of the information present within the original "C" or "Ku" satellite frequency bands. The IF output of the LNB typically ranges from 950 to 1450 MHz (950 to 2050 MHz for units now sold in Europe and elsewhere overseas). The block IF signal is then filtered and amplified before being sent to the indoor unit via a length of coaxial cable.

Every point discussed so far is common to both the new digital TV systems and older satellite TV transmission standards that have been around now for more than a decade. Before we delve into the realm of digital satellite technology, let's examine the less complex world of analog satellite TV. The signals are called "analog" because they vary continuously in frequency and intensity, whereas digi-

Fig. 3-2. The stages of signal processing in an analog satellite TV receiving system. Above all, don't panic! Some of the more esoteric items will be explained in due course. For maximum immediate effect, however, leave this book out on the coffee table so that it lies open to this page. That way your friends will all think you are an industrial spy!

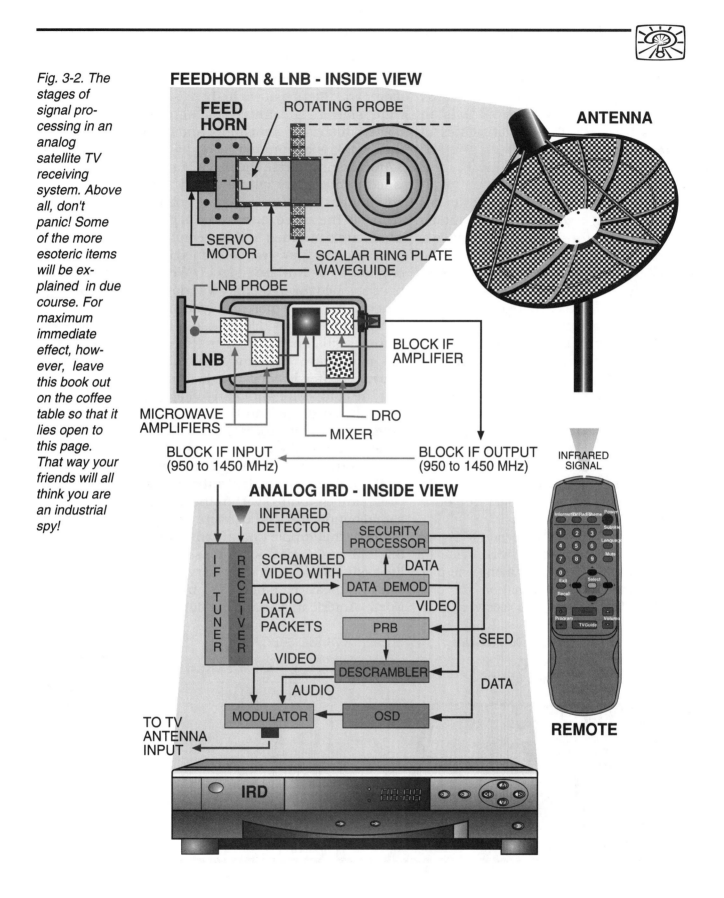

FEEDHORN & LNB - INSIDE VIEW

FEED HORN
ROTATING PROBE
ANTENNA
SERVO MOTOR
SCALAR RING PLATE
WAVEGUIDE
LNB PROBE
LNB
BLOCK IF AMPLIFIER
MICROWAVE AMPLIFIERS
DRO
MIXER
BLOCK IF INPUT (950 to 1450 MHz)
BLOCK IF OUTPUT (950 to 1450 MHz)
INFRARED SIGNAL

ANALOG IRD - INSIDE VIEW

INFRARED DETECTOR
SECURITY PROCESSOR
IF TUNER
RECEIVER
SCRAMBLED VIDEO WITH
DATA
AUDIO DATA PACKETS
DATA DEMOD
VIDEO
PRB
SEED
VIDEO
DESCRAMBLER
DATA
AUDIO
MODULATOR
OSD
TO TV ANTENNA INPUT
IRD
REMOTE

tal signals convert the sound and picture information into strings of zeros and ones—binary digits called "bits" to represent the "on" and "off" states of computer logic circuitry.

THE ANALOG IRD

Inside the analog IRD, another mixer circuit, which is controlled by the unit's channel selector, is used to produce a second IF signal that contains a single satellite channel. When we say second IF, we actually mean a band of frequencies centered on a particular frequency. Some manufacturers use a 70-MHz center frequency for their second IF signals while others may use 130-, 140-, or even 510-MHz. Some analog set top boxes provide external loop-through connections to the second IF portion of the circuitry so that filters can be added to enhance reception or suppress terrestrial interference. Other more advanced units contain built-in filters that can be selected by the operator on a channel-by-channel basis.

The bandwidth of the second IF signal needs to be around 27 MHz wide in order to effectively pass along the picture and sound information for one analog TV channel. The receiver takes the second IF signal, amplifies it some more, and filters it down to the 27-MHz-wide signal. Unwanted background noise, which is present above and below the signal, is eliminated.

After being filtered, the block IF signal is fed into another circuit called a discriminator. The discriminator's job is to strip off the video and audio signals, which are contained in the second IF in the same way they were contained in the original C- or Ku-band signal. We may have stepped the frequency way down in order to handle it more easily, but that doesn't affect the video or audio information at all. For this reason, the original microwave frequency is called a "carrier." After leaving the discriminator, the composite audio and video signals are separated and passed on to other circuits for processing.

Nearly every unencrypted analog satellite TV transmission uses one or more additional carriers, called "subcarriers," which are contained within the 27 MHz signal bandwidth. Subcarriers are used to carry the supporting audio (usually 6.2 and/or 6.8 MHz for domestic satellite systems) for each video service. The stereo subcarrier pairs used by many North American C-band satellites are

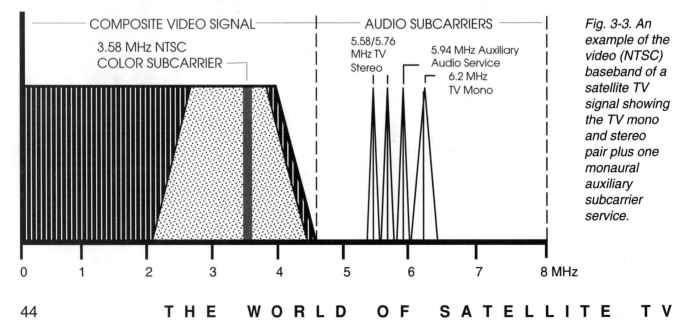

Fig. 3-3. An example of the video (NTSC) baseband of a satellite TV signal showing the TV mono and stereo pair plus one monaural auxiliary subcarrier service.

located 5.58/5.76, 5.94/6.12, 6.3/6.48, 7.02/7.20, 7.38/7.56, and 8.10/8.28 MHz above the center frequency of the transponder's main carrier.

The analog set-top box provides a control that allows the operator to select the appropriate subcarriers for the desired channels. On some satellites there also may be several auxiliary subcarriers attached to any one satellite TV channel. The latest analog-based IRDs either come pre-programmed from the factory with all the subcarrier frequencies stored in memory or can automatically scan for new satellite audio services.

The extra audio subcarriers are used to transmit broadcast radio programs, in-store music for shopping malls, or carry data transmissions. They do not interfere at all with the video program and offer a way to provide multiple use for these expensive satellite transponders.

All satellite set top boxes these days come with a wireless remote control for channel selection. Most models also provide an on-screen graphics display that will tell you which satellite channel you are viewing. Other channel selection features include channel up/down, direct access, scan video and audio tuning, designated "favorite" video and audio channels, and automatic picture fine-tuning.

UNSCRAMBLING THE SIGNALS

Encryption is an electronic means of rendering the video and audio reception of a program unintelligible except to those stations that can reassemble the signals. All other stations intercepting any "scrambled" analog TV program service will encounter distorted video and no audio, or perhaps in some cases a "barker channel" that provides an audio message about how to subscribe.

The picture is commonly secured by converting the video's standard synchronization codes to an alternate format. This distorts the video in such a way that a standard TV set cannot correctly interpret the signal. For example, the encryption system may turn the picture inside out, thereby creating a kind of negative image. Other systems slice and dice the video images and then randomly rearrange the segments into colorful patterns that may be of interest to a few aging flower children from the 60s, but otherwise are not an entertainment option. If there is a category in the *Guinness Book of Records* for the number of hours spent staring at these images, I think that I just might qualify for the world record.

The slice and dice method, which is more formally referred to as the line rotation or "cut and rotate" method, uses an electronic "key" called an algo-

Fig. 3-4 & 3-5. Typical displays of encrypted analog satellite TV signals.

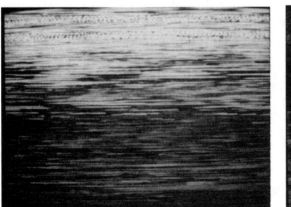 

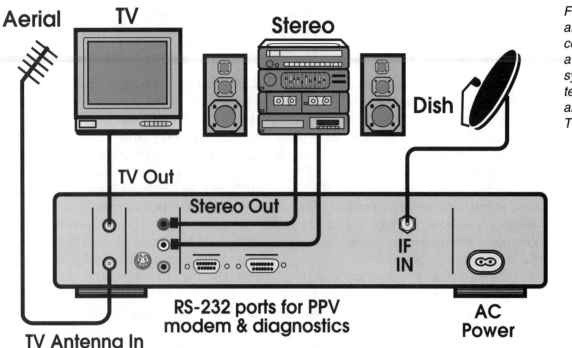

Fig. 3-6. IRD auxiliary connections to a home stereo system, terrestrial TV antenna and TV set.

rithm to select the locations of the cut points. The mathematical algorithm also controls the rearrangement of the line segments back to their original places. Unless the decoder has access to this key, the picture remains an unintelligible jumble.

The sound portion of the programming is commonly changed from an analog waveform to a digital bit stream. Like the conversion of analog-based tapes and records into digital CDs, this digital conversion process actually improves the fidelity of the audio signals. The digital bit stream also is usually encrypted to prevent unauthorized reception.

Each program service provider using an encryption system can individually address every subscriber by means of an authorization code that is transmitted at regular intervals over the satellite to every system receiving the scrambled signals. This code is used to switch on (or to switch off if you don't pay your bill) each decoder's access to the required electronic keys. Since any subscriber's decoder can be turned on or off within a few minutes, this addressable technology also makes it possible for some programmers to offer special pay-per-view movies and other special events. In this case, the addressable code is sent over the satellite to the IRD's decoder module at the beginning of a selected pay-per-event and then removed from the satellite signal at the conclusion of the program.

At the heart of most encryption systems is the subscriber's security card, a rectangular piece of plastic about the size of a credit card which stores special codes which the IRD must have to complete the decoding process. The codes reside inside a solid-state electronic chip that is embedded within the IRD's "smart card." To activate the IRD, the smart card must be inserted into the conditional access (CA) module slot which is usually found on the front panel of each IRD.

The smart card may be authorized to

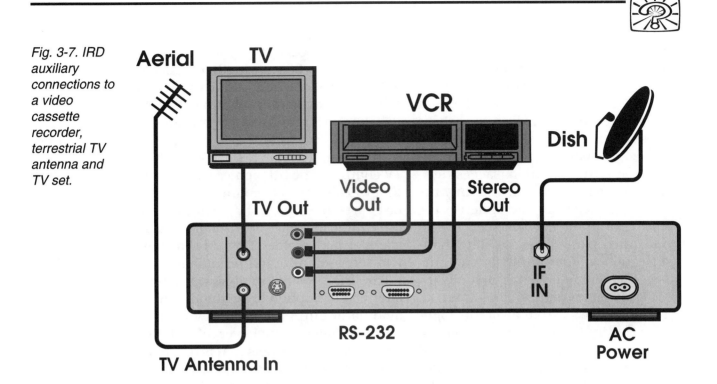

Fig. 3-7. IRD auxiliary connections to a video cassette recorder, terrestrial TV antenna and TV set.

receive a single program service or an entire package or tier of services, depending on the subscription or subscriptions that the subscriber has elected to purchase. If at any time the programmer's security system is compromised, the service provider can switch to a new set of codes. In this case, the programmer will be required to send its subscribers new smart cards containing the new coding information.

RECEIVING DIGITAL TV SERVICES

Many programmers are now using digital video compression to fit multiple TV signals onto one satellite channel or transponder. A sixteen-transponder satellite that formerly could only offer a maximum of sixteen analog TV channels can now deliver more than one hundred digital TV services. This economy of scale has allowed programmers such as HBO and Showtime to offer a different version of their movie services for each time zone. At any given time of the day or night you can have five

or more movie selections per service from which to choose.

With digital video compression, each program's video and audio components are converted from a standard analog waveform into a stream of numerical digits. Compression is achieved by using abbreviated numerical codes as well as by employing transmission techniques that only transmit the portion of the TV picture that actually changes from one complete image or "frame" of video to the next.

Multiple TV and audio channels are combined or "multiplexed" into a single digital bit stream that contains the video and audio programming for each available service. It also includes the electronic keys that your IRD needs to decode the channels which you have been authorized to receive. The same digital bit stream also contains the programmer's EPG and special service information that will automatically reconfigure the IRD in the event that the programmer changes any of its main transmission parameters.

If the digital IRD has not been authorized to access a digitally compressed service, it will display a blacked-out screen containing a message such as "no conditional access" or "smart card required." If the set top box is not compatible with the conditional access system that the programmer is using, the decoder may not be able to detect the presence of any signal whatsoever, even when tuned to the correct satellite channel frequency. For all intents and purposes, the digital signal is virtually indistinguishable from random noise.

The good news is that all of this high-tech wizardry is totally transparent to the TV viewer. To receive a standard broadcast TV service, you need to know the channel on which a particular TV program will appear. With a digital IRD, however, you can easily find the desired service in a list of favorite channels or use the EPG until you find something you might like to watch. Just click a button on your remote and the program pops onto your TV screen. If a favorite program service changes its channel frequency, you will not even notice. The service information within the digital bit stream entering the IRD will automatically reprogram the unit for you.

There's also no such thing as a fuzzy digital TV picture, so there are no filters or fine-tune controls that require adjustment. If you are one of those people who enjoy fiddling around with consumer electronic gear, don't despair. There are some things which you can do to customize your digital IRD, such as establish one or more lists of favorite channels, establish program ratings for your youngsters, or set spending limits for accessing pay-per-view movie services. The details on these and other extracurricular activities will be presented a little later in this book.

THE RF MODULATOR

In all set top boxes, whether analog or digital, the video and audio signals are connected to an internal unit called the RF (short for "radio frequency") modulator. The RF modulator acts as a miniature transmitter for a selected regular TV channel. The IRD comes with a cable that will allow you to connect the set top box directly to your TV set's antenna input, if it has one, or through a CATV transformer to older TV sets that have two screw-on antenna terminals.

The IRD's RF modulator will usually offer you the option of switching the RF modulator output to either TV channel 3 or 4. Select whichever of the two terrestrial TV channels that is not in use at your location. This will eliminate interference to your satellite TV viewing caused by a strong local TV channel. This will also permit you to continue to receive all the local TV channels that are broadcast over the air in your area.

The IRD also provides a video output port as well as right and left audio output connections on its back panel. Some units also feature an "S" video output for connecting to other compatible video electronic components such as a VCR or camcorder. The direct video and audio connections will deliver signals that are of a superior quality to what you can obtain when you connect an RF modulator to the TV set's antenna input.

Many satellite-delivered entertainment services offer their programs in stereo. Encrypted services may even offer digital, compact disc quality stereo with surround sound. Whenever superior audio reception is desired, the IRD's audio outputs should be connected directly to the "left" and "right" audio input connectors located on the back panel of a home entertainment sound center or a stereo-ready TV set.

SATELLITE ANTENNAS

Around the world these days, satellite TV antennas are sprouting up like mushrooms. Some of these dishes are huge; others, no bigger than a pie tin. The rest are sized somewhere in between. This gives rise to the question: "Why is there such a large variation in the size of the dish that people use to receive satellite TV signals?" As we discovered in the previous chapter, the consumer selects the antenna size that is required to receive services from one or more satellites. In this chapter

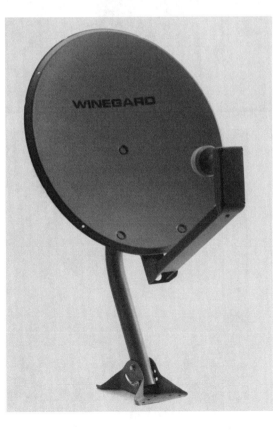

Fig. 4-1. This 76cm offset antenna for Ku-band DTH operations is constructed of galvannealed steel with a textured powder coat finish. (Courtesy Winegard.)

we will examine the technical reasons for determining antenna size as well as how the various types of antennas and dish mounts operate.

KU-BAND ANTENNA SIZE LIMITATIONS

Most Ku-band satellites serving the Americas can deliver DTH TV services within their respective coverage areas to antennas ranging from 1.5 to 3 feet (45 to 90cm) in diameter. One primary determinant of the required antenna size is the amount of power—typically 45 to 240 watts per transponder—that each Ku-band satellite can deliver. Another factor is the size of the satellite's coverage area or footprint. A 50-watt satellite signal that is tightly focused into a relatively small coverage area can produce the same signal strength on the ground as a 120-watt satellite with a much larger footprint.

Signal strength isn't the only consideration in determining the size of a Ku-band dish. A "parabolic" curve is typically used in the design of all satellite dishes. One important property of this parabolic curve is that it creates a narrow beam through which the dish looks up at the sky—the smaller the antenna the wider its receiving beam. Ku-band DTH satellites using the 11.7 to 12.2 GHz spectrum operate in a closely spaced orbital environment. Other satellites using the same frequency spectrum may be located just two degrees to either side of the DTH satellite. The 45cm antenna

Fig. 4-2. A C-band satellite antenna farm installed at one of the early U.S. trade shows.

Fig. 4-3. A 3m C-band satellite dish equipped with a motorized antenna actuator. (Courtesy Orbitron.)

Fig. 4-4. In 1982, I photographed this DX Antenna display of a complete Ku-band DTH system that, in many re-spects, was a foreshadowing of the DBS era to come.

has a broad receiving beam that can see up to three Ku-band satellites at the same time in this environment. That is why some Ku-band DTH services use 60 or even 90cm dishes with narrower receiving beams so that potential interference problems between similar closely-spaced satellites can be avoided.

High-power Ku-band DBS satellites do not have this problem because the RARC-83 DBS plan for the Americas stipulates a minimum separation of nine degrees between satellites serving the same area or adjacent coverage areas. For example, the DBS orbital assignments for the United States are located at 61.5, 101, 110, 119, 149, 158 and 166 degrees west longitude.

C-BAND ANTENNA SIZE LIMITATIONS

C-band DTH dishes are almost always far larger than their Ku-band counterparts. The C-band satellites serving the Americas typically transmit using just 16 to 35 watts of power per transponder. The lower power levels are required to prevent interference to ground-based telephone microwave links and other telecommunication services sharing the C-band frequency spectrum.

The relatively broad corridor along which the C-band antenna looks up at the sky is another limiting factor. The "beamwidth" of any parabolic dish is a direct function of the frequency that the antenna is receiving—the lower the frequency the wider the receiving beam for any given dish size. For example, a 60cm antenna would have a beamwidth of almost 9 degrees when used to receive a C-band signal, and a beamwidth of about 3 degrees when receiving a Ku-band signal.

In some instances, a 1.2m C-band antenna could produce a perfect TV picture from one or more of the medium-power C-band satellites serving

the Americas. However, the 1.2m dish has a C-band antenna beamwidth of more than 4 degrees, which means that signals coming from an adjacent C-band satellite just two degrees away would also fall within the antenna's main beam. For this reason, a minimum antenna size of 1.8m is recommended for most C-band DTH system installations.

SMALL APERTURE KU-BAND DISHES

The satellite receiving dish captures the intended satellite signal and reflects it back to a common point called the focal point. The antenna's aperture collects a relatively weak signal spread across the reflector surface and concentrates it, making it stronger.

The performance of any type of satellite dish is primarily governed by the accuracy of its curvature. No matter what materials are used, the dish is only as accurate as the mold and dies used to produce it. For this you will have to depend on the reputation of the manufacturer.

Manufacturers can use a wide variety of materials to construct their antennas. Solid aluminum dishes are rugged and tend to hold their exact shape over the lifetime of the system. Aluminum antennas usually are coated with a special non-reflective paint to keep sunlight from being reflected to the focal point where it could cause damage to other system components. This type of antenna usually is constructed from marine-grade aluminum that prevents corrosion.

A solid antenna with a bunch of holes drilled into it to lessen its environmental impact is called a perforated antenna. The diameter of these holes, which must be relatively small in comparison to the satellite signal's wavelength, presents a totally reflective surface to the incoming microwaves.

Fiberglass dishes, if well constructed, can be as good as the solid aluminum type and should work fine for all DTH receiving applications. Most standard fiberglass dish antennas have a wire mesh embedded inside. Without a metal reflector, the microwave signals would pass straight through the fiberglass. The best fiberglass dishes are pressure-molded from fiberglass-reinforced plastic.

Offset-fed antennas. Most Ku-band DTH and DBS receiving systems use the offset-fed antenna design. The offset-fed antenna consists of a segment of a complete parabolic curve to create a focal point that lies below the center of the reflector. One important advantage of the offset antenna is that it doesn't have

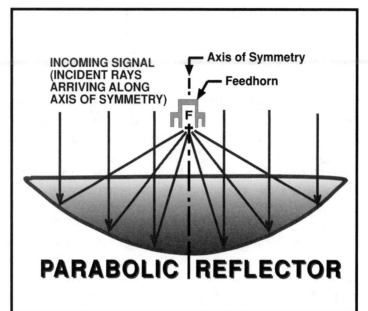

Fig. 4-5. The basic principle of the parabolic reflector is that all incident rays of energy arriving along the antenna's axis of symmetry are reflected to a common focus.

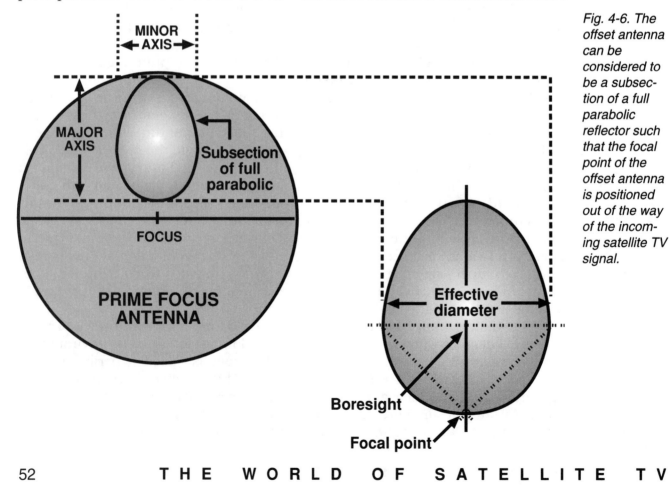

Fig. 4-6. The offset antenna can be considered to be a subsection of a full parabolic reflector such that the focal point of the offset antenna is positioned out of the way of the incoming satellite TV signal.

to tilt upwards very high to receive any given satellite. Rain and snow therefore tend to slide off of the reflector's surface. Another advantage is that the LNB and feedhorn are positioned below the front of the reflector, outside of the signal pathway between the antenna and the satellite. Since the LNB and feedhorn do not block the incoming signal, the performance efficiency of the dish is slightly enhanced.

Offset-fed antennas are oval rather than circular in shape, with a "major axis" and a "minor axis". The antenna can be installed in such a way as to reduce the amount of signal gain in the east/west plane, which reduces the amount of interference that the dish might otherwise receive from adjacent Ku-band satellites in a closely spaced orbital environment.

The offset design approach also minimizes any noise contributions from the surrounding environment. The feedhorn tilts up to look at the sky rather than down at the Earth, which generates a much higher noise level.

Flat antennas. Alternative antenna designs, such as the early Japanese prototype shown, are being used in Japan and elsewhere for receiving TV signals from high-power DBS satellites. Instead of using a parabolic shaped reflector to concentrate the incoming signals, the flat antenna, called a "planar array," uses an embedded gridwork of tiny resonant elements that receives the satellite signals and concentrates them at a common terminal point at the rear where the LNB is mounted.

One advantage of the flat plate antenna is that it is relatively unobtrusive. No feedhorn is required, and the LNB is located out of sight directly behind the plate. This type of antenna is usually mounted on an outside wall or a rooftop. Flat antennas, however, are almost always dedicated to the reception of a single satellite or multiple satellites collocated around a single orbital position. The cost of manufacturing flat antennas is higher than what it costs to build parabolic dishes. As a consequence they are not widely used in the Americas to receive satellite TV signals.

LARGE APERTURE C-BAND ANTENNAS

Most C-band receiving dishes use a "prime focus" design where the focal point of the parabolic antenna is located directly over the front and center of the dish. The feedhorn and LNB assembly mount directly above the parabolic dish, centered at the focal point of the parabolic curve for maximum signal pickup.

Most feedhorn/LNB assemblies on a prime focus antenna use a multi-legged support bracket that bolts onto the surface of the dish. This support bracket positions the assembly over the center of the dish at the required distance. A sturdy support structure is important here, since being off an inch or more can

Fig. 4-7. In 1985, I was invited to a demonstration of new antenna technology developed by Matsushita Electric Works in Osaka. The photo shows a technician holding a 15-inch flat antenna near a window to receive TV from the Yuri DBS satellite.

be critical. Another type of LNB/feed-horn support is the buttonhook, which uses only one support member to position the LNB and feedhorn.

An expanded aluminum mesh is currently the most popular material used to construct the reflector surface of the dish. It has a low environmental impact and also can be readily shipped in a relatively small package by delivery services such as UPS and DHL. Since you can see through the dish, it stands the least chance of drawing complaints from neighbors in crowded urban and suburban areas.

The gaps in the mesh material are small enough to present what appears to be a solid reflective surface to the incoming C-band satellite signals. Ku-band signals have a much shorter "wavelength" (2.5cm) than C-band (7.5cm) signals. If you intend to use a mesh dish to receive Ku-band satellite signals be sure that the mesh material is rated for Ku-band use prior to purchasing it.

Mesh dishes are perhaps the most susceptible to damage from the elements. A careful inspection of the method used for fastening the screen to the support members is advisable. You can also check a mesh dish for curvature inaccuracies by running your hand across its surface. Noticeable bumps or waves in the surface material indicate deviations from the precise parabolic curve necessary for good performance.

Solid aluminum and fiberglass models are also available, which may come in a single piece or as several petals that bolt together. The one-piece dish conforms most accurately to the manufacturer's master mold but is a problem to ship. The petal dishes are stronger, easier to transport, and one or more petals can be replaced if damaged. Aluminum reflectors may also be perforated to lessen their environmental im-

pact. Again, the manufacturer selects the diameter of the perforations with the wavelength (7.5 versus 2.5cm) of the incoming C- or Ku-band signal in mind.

When examining any dish you should look at how the various petals fit together. Does the surface look continuous, with little variation from petal to petal? Keep in mind that variations in curvature of as little as a quarter of an inch can affect the performance of your system.

Stand to one side and look across the surface of the dish to visually detect any warp to the antenna's parabolic curve. The outside edges should line up with each other. You can also detect antenna construction deviations by attaching two or more strings to the rim of the antenna and passing them across the face of the dish so that they bisect the center of the reflector. The strings should all lightly touch at the point where they cross. If there are gaps between the strings, this most always is an indication that the antenna was improperly installed. Poor quality control at the factory also will occasionally result in the release of a product that does not conform to the manufacturer's design specifications.

SPHERICAL & CASSEGRAIN ANTENNAS

While there are other types of antennas that are used for satellite TV reception, their use is primarily limited to cable and SMATV (Satellite Master Antenna Television) applications. The two alternative designs seen most often are the spherical and the cassegrain antenna.

The perimeter of a spherical antenna is usually square or rectangular in shape. The spherical antenna has a curvature that, if extended outward, would become a sphere. It also has the property of creating multiple focal points so that two or more satellites can be simultaneously received without the need to move the dish.

Installed in a permanent, fixed position, the spherical antenna can simultaneously capture the signals from satellites located across a wide section of the sky. Each feedhorn/LNB assembly mounts on an adjustable support located at the front of the antenna. Multiple satellites are received by positioning several assemblies at the various focal points generated by the spherical antenna. The range of reception is approximately 20 degrees to either side of the antenna's axis of symmetry.

The cassegrain antenna uses a small subreflector located at the focal point of the dish to reflect the gathered signal back to the center of the reflector where a rectangular feedhorn is mounted. The LNB attaches to a feedhorn flange lo-

cated at the rear of the dish which makes servicing an easier proposition. This design approach also minimizes the noise contribution of the surrounding environment. The feed is located inside the dish and looks up at the sky rather than down at the Earth. This dramatically improves the antenna's ability to reject noise that might otherwise degrade the signal level produced by the receiving system.

The cassegrain design, which typically is used for large aperture antennas with a diameter of 5m or larger, achieves a slightly higher G/T performance level than a prime focus antenna of equivalent size. Smaller antennas, however, cannot use this design approach because the large subreflector required would block too high a percentage of the signal from reaching the antenna's reflector, thereby negating any performance gains that might otherwise have

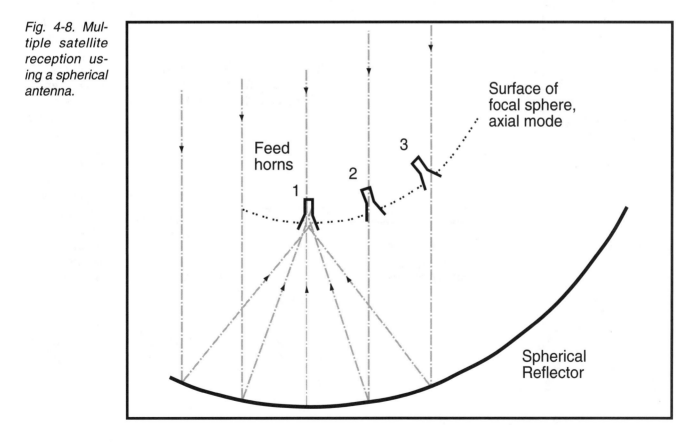

Fig. 4-8. Multiple satellite reception using a spherical antenna.

been achieved.

ANTENNA GAIN

The performance of a dish antenna is most commonly rated in terms of gain—a measurement of the passive (nonelectronic) amplification that a dish gives to the signal that it receives. Quite simply put, the bigger the dish, the larger the capture area and the higher the gain. Gain is expressed in decibels, or dB.

Dish gain measurements are made in reference to a standard isotropic antenna—an imaginary aerial conceived by engineers which receives signals equally from all directions at once and thus has a gain of 0 dB. Real-life antennas are measured by determining how many times better they perform in a given direction than the isotropic standard.

The curvature of the dish acts much like a lens to concentrate the signal within a narrow corridor called the "main beam." This area lies directly to the front and center of the prime focus antenna and is obtained through maximizing the signal present in the antenna's main beam while limiting signals or noise coming from other sources and directions. The amount of signal delivered by the dish is critical since the system's first stage of electronic amplification can only amplify as much signal as it receives.

3 dB = a doubling of power;
10 dB = x 10;
20 dB = x 100;
30 dB = x 1,000;
40 dB = x 10,000;

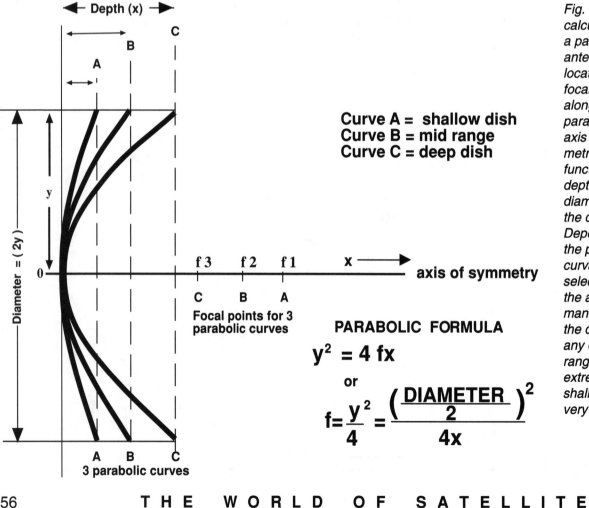

Curve A = shallow dish
Curve B = mid range
Curve C = deep dish

PARABOLIC FORMULA

$$y^2 = 4\,fx$$

or

$$f = \frac{y^2}{4} = \frac{\left(\dfrac{DIAMETER}{2}\right)^2}{4x}$$

Fig. 4-9. The calculation of a parabolic antenna. The location of the focal point along the parabola's axis of symmetry is a function of the depth and the diameter of the dish. Depending on the parabolic curvature selected by the antenna manufacturer, the depth of any dish can range from extremely shallow to very deep.

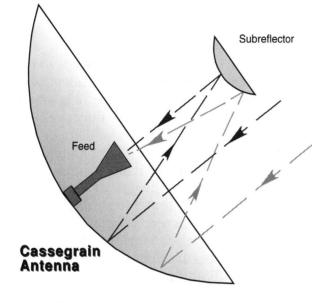

Fig. 4-10. Line drawing showing the geometry of a cassegrain antenna.

Subreflector

Feed

Cassegrain Antenna

50 dB = x 100,000.

A chart is included so that you can determine typical gain figures for a variety of dish sizes and satellite frequencies. From this chart we also can see that the gain of the antenna varies according to the frequency being received. But don't be deceived. Any increase in dish gain as the receiving frequency rises is offset by a corresponding increase in the losses that occur as the higher-frequency signal travels from the satellite to the surface of the Earth below.

Expressed as a percentage, the efficiency rating of the dish represents that portion of the signal striking the antenna's surface that actually arrives at the feedhorn. Efficiency ratings for prime focus antennas typically range from 55 to 65 percent, while offset-fed and cassegrain antennas can achieve even higher efficiencies from 65 to 75 percent. Manufacturers optimize gain performance by improving the dish's efficiency either through the use of tighter curvature tolerances or alternate feed techniques. The real trick, however, is

Fig. 4-11. Nominal antenna gain versus frequency chart.

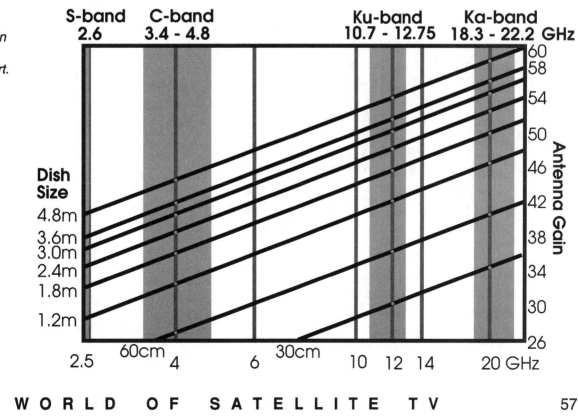

to "illuminate" the antenna in such a way that the ratio of gain to system noise, called the G/T (pronounced "Gee over Tee"), achieves the maximum possible value.

ANTENNA DEPTH

Depending on the parameters of the parabolic curve selected by the antenna designer (see Fig. 4-9), the depth of any dish—the distance from a point at dish center in line with the antenna's rim and the center of the reflector surface—can range from extremely shallow to very deep. A shallow dish will have a very long focal distance, so the feedhorn and LNB will be positioned a considerable distance from the center of the reflector. The shallow dish design maximizes the potential gain of the antenna since the feedhorn has an excellent view of the entire reflector.

There are several disadvantages to the shallow dish design. The molecular motion of the Earth itself generates a random noise pattern that permeates the entire electromagnetic spectrum, including the satellite frequency bands. This noise combines with the incoming satellite signal and lowers the system G/T, the ultimate figure of merit for judging system performance.

The primary source of noise, expressed as a "noise temperature," is the Earth itself. As the shallow dish tilts downward to receive satellites that are relatively close to the horizon, the antenna also receives the "hot" noise temperature radiated by the Earth.

When used to receive C-band satellite signals, the shallow dish is also potentially vulnerable to terrestrial interference from nearby microwave relay stations. The deep-dish design positions the LNB and feedhorn at a level almost parallel with the rim of the antenna. This permits the deep dish to provide

some shielding for the feedhorn so that less noise and terrestrial interference is able to gain entry to the system. Because of the feed's close proximity to the surface of the antenna, it cannot view or "illuminate" the entire reflector surface.

The deep dish typically may produce less gain than a shallow dish of the same diameter while at the same time producing a higher gain-to-noise ratio or G/T. The deep dish is the best solution for receiving satellites that are relatively low to the receiving site's horizon or when potential terrestrial interference sources are operating in the vicinity of a C-band installation.

THE F/D RATIO

The relative depth or shallowness of the dish is often expressed as a focal length to diameter (f/D) ratio. The focal length is the distance between the lip of the feedhorn's central opening and the center of the surface of the dish. The depth of the dish is the distance from a point at dish center in line with the antenna's rim and the center of the reflector surface. Prime focus deep dishes have f/D ratios of .25 to .33,

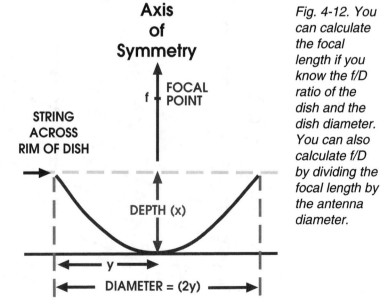

Fig. 4-12. You can calculate the focal length if you know the f/D ratio of the dish and the dish diameter. You can also calculate f/D by dividing the focal length by the antenna diameter.

while the f/D of shallow dishes will range between .4 and .5. If you do not know what the focal length should be, you can calculate it by multiplying the diameter of the antenna by its f/D ratio.

ANTENNA SIDE LOBE REJECTION

The perfect parabolic antenna would only amplify the satellite that is located directly to the front and center of the dish while totally rejecting signals and noise sources coming from all other directions. In reality, every antenna has its own unique receiving pattern that illustrates how much gain will be produced from signals emanating from a given direction.

When a manufacturer specifies the gain of his antenna, he is referring to the amplification provided by the main beam of the antenna's receiving pattern. However, all antenna radiation patterns also reveal the presence of multiple "sidelobes" that will amplify signals coming from angles adjacent to the direction of the antenna's main beam. For most digital DTH applications, the sidelobe gain should be at least -12 to -15 dB below that of the main beam.

Only the high-power DBS satellites have the protection afforded by a minimum of nine degrees of orbital separation between DBS satellites serving the same region. In all other cases, two-degree orbital spacing between satellites is typically an operational reality. Moreover, newer satellites are transmitting at higher power levels than ever before. Both of these developments make antenna sidelobe performance an important consideration.

DTH service providers minimize the problem of adjacent satellite interference by specifying antennas that produce a main beam that focuses on a single satellite and drop the signals

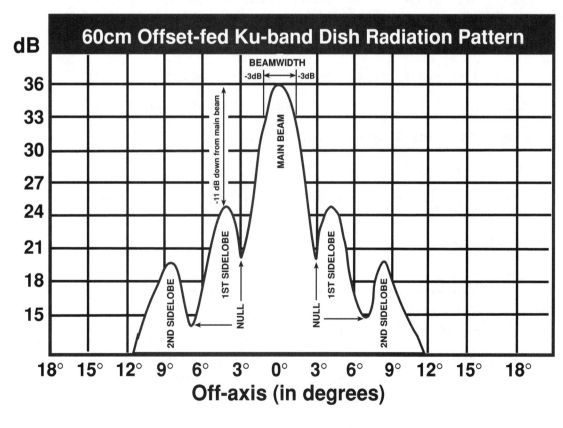

Fig. 4-13. This antenna radiation pattern for a 60cm offset antenna shows less than -12 dB attenuation of the first sidelobe.

Construction:	4 piece preassembled, 12 aluminium ribs, Ku-band mesh, powder coat finish
Diameter (in feet/meters):	10 ft/3m
Frequency Range:	3.7 to 4.2 GHz (C-band DTH) 11.7 to 12.2 GHz (medium power DTH) 12.2 to 12.7 GHz (high power DBS)
Antenna Gain in dB: at 4 GHz: at 12 GHz:	 40.5 dB at 65% efficiency 49 dB at 55% efficiency

(The gain of any antenna will vary slightly depending on the frequency; the higher the frequency, the higher the gain. See chart previously presented.)

Feed:	Prime focus, quad support
f/D Ratio:	.375 *(focal length to antenna diameter ratio)*
Focal Point:	45"
Side Lobes:	20dB down
Operational Winds:	60 mph

(how high a wind speed the dish can operate under without significant picture degradation)

Survival Winds:	120 mph

(the speed beyond which damage to the dish will be likely to occur)

Mount:	Horizon-to-Horizon Modified Polar
Azimuth Tracking Range:	180° Azimuth adjust
Elevation Adjust:	5° to 90°

(the dish can be pointed at any satellite that lies within these elevations at your location)

Pole Size Required:	3.5" O.D.
Shipping Weight:	160 lbs.

Fig. 4-14. Typical specifications sheet for a 3m prime focus satellite dish.

from adjacent satellites into the first "null" of the antenna receiving pattern.

MOUNTS

The mount is often the most overlooked aspect of any parabolic antenna. The steel mount and the support bearing which holds the dish are very important since they not only carry the full weight of the dish but must also maintain a precise position once sighted onto a satellite. Changes in alignment

Fig. 4-15.
Typical
specifications
sheet for a
range of Ku-
band offset
antennas.

DISH DIAMETER:	40cm	60cm	70cm	85cm
CONSTRUCTION:	Steel or Aluminium with powder coat finish			
BANDWIDTH:	10.7 - 12.75 GHz ALL			
GAIN (@ 12.5 GHz):	33 dB	36.1 dB	37.5 dB	38.9 dB

(The gain of any antenna will vary slightly depending on the frequency; the higher the frequency, the higher the gain. See chart previously presented.)

EFFICIENCY:	70 PERCENT ALL			
BEAMWIDTH (-3 dB):	4.3°	3.1°	2.4°	2.3°
f/D RATIO:	.6 ALL			

(Note that offset antennas typically use an f/D of .6 to .7 as opposed to prime focus parabolic antennas that use an f/D of .25 to .5. These extremely shallow dishes appear almost flat instead of bowl shaped.)

OFFSET ANGLE:	26.0°	22.6°	26.0°	21.3°

(This is an important specification for installers to know. Most of the charts and computer programs used to calculate antenna elevation angles are for prime focus antennas, not offset antennas. To derive the correct angle from one of these charts or computer programs, the installer must subtract the offset angle from the value provided.)

ELEVATION ADJUST:	10 - 50° ALL			

(The dish can be pointed at any satellite that lies within these elevations at your location.)

SURVIVAL WINDS:	130 km/hr ALL			

(The speed beyond which damage to the dish will be likely to occur.)

DISH WEIGHT:	2.1 kg	3kg	3kg	4.9kg
MOUNT OPTIONS:	Wall, rooftop, pole and terrace mounts available.			

as small as a couple of inches can make the difference between a good picture and no picture at all. One good test for any antenna's mount is to grab the rim of the dish and shake it. If the dish support has a significant amount of looseness or "play" in it, the wind and rain will be able to shove the dish about

which can cause erratic
Since the height or "
sky and the direct
each satellite wil'
must incorp'
its structu'
focused on
There are'

used to support parabolic dishes: the fixed, elevation over azimuth (El/Az) and modified polar.

FIXED AND EL/AZ MOUNTS

The fixed mount is commonly used by all digital DTH and DBS systems designed to receive a single satellite or constellation of satellites clustered about a single orbital location. During the installation, the fixed mount requires the independent alignment of two mount adjustments called the elevation and the azimuth.

The elevation is the angle at which the dish tilts up toward the sky. The azimuth is the direction or "bearing" from the site to a point on the horizon that intersects with a vertical line extending down from the satellite's location in the sky. These alignment adjustments are locked in place once the antenna has been peaked to achieve maximum signal level.

Commonly encountered in commercial rather than home satellite TV systems, the El/Az mount is a motorized version of the fixed mount that requires the independent adjustment of both the elevation and azimuth in order to receive any particular satellite. Since it requires two adjustments to switch satellites, it is not easy to manually maneuver. Computerized controls are usually employed at commercial installations to make any required satellite changes.

MODIFIED POLAR MOUNTS

Astronomers have long used the polar mount because it allows their large telescopes to track distant stars and galaxies as the Earth rotates. A modified version of the polar mount is used by all motorized DTH receiving systems. The dish antenna rotates in an arc that mimics the curvature of the Clarke Orbit. The axis of the polar mount must be installed so that it faces true south (true north for stations south of the equator). The true north/south line for any site location is calculated by using a compass to determine the location of magnetic north and then adding or subtracting a correction factor obtained from available maps or your local airport control tower. Once the antenna and mount have been properly installed, the polar mount antenna can track all other satellites within view of the dish by means of a single adjustment in direction (called the azimuth). The modified polar mount will automatically change the antenna's elevation angle as the dish moves to the east or west.

For exact tracking of the Clarke Orbit from the eastern to the western horizons, a modified polar mount includes an additional adjustment called the declination. The amount of declination required at any location will vary, and it must be set correctly during the initial installation of the dish and mount in order for the antenna to properly track the entire portion of the satellite arc that is visible from the site location. All the details concerning how to install satellite antennas are provided in the Installation chapter.

THE FEEDHORN, LNB & LNF

Before we can understand the performance characteristics of any satellite TV system component, we must first know the identity and true nature of our principal enemy: noise. Molecular motion within all matter generates a band of random noise that permeates almost the entire electromagnetic spectrum, including the frequency bands used for satellite communications.

The incidence of noise increases as the temperature of the matter involved also increases. Therefore, the noise generated by molecular action is called "thermal noise," the level of which can be expressed as a noise temperature or an equivalent noise figure.

NOISE TEMPERATURE

The temperature scale used to quantify the level of any thermal noise source uses measurement units of degrees Kelvin (K). The Kelvin temperature scale can be related to more commonly understood temperature scales in Celsius or Fahrenheit. The noise temperature at which all molecular motion stops is called "absolute zero" or 0 K. Physicists and cosmologists have shown, however, that "absolute zero" does not exist in nature. Cosmologist Stephen Hawking discovered that the thermal noise temperature of deep space even exceeds absolute zero by a fraction of a degree.

The Earth, which has a noise temperature of 290 K, is a "hot" noise

Fig. 5-1. Comparison chart between the Celsius, Fahrenheit and Kelvin temperature scales.

TEMPERATURE COMPARISONS

Thermal noise is measured in degrees Kelvin

Absolute zero: temperature at which all molecular motion stops.

	Kelvin	Celsius	Fahrenheit
Steam	373°	100°	212°
Earth	290°		63°
Ice	273°	0°	32°
Solid CO$_2$	195°	-98°	-109°
Liquid Oxygen	-90°	-183°	-297°
Absolute Zero		-273°	-460°

source. Radiation from the Sun warms the Earth, which excites molecular motion. The Sun itself generates a noise temperature of 10,000 K, while other heavenly bodies have unique noise temperatures of their own.

The cumulative amount of thermal noise that surrounds us is many times stronger than the satellite signals reaching the site location. The task of the feedhorn, located at the focus of the receiving dish, is to efficiently gather the signal captured by the reflector, while at the same time minimizing any outside noise contributions.

The dish may be the most visible portion of any satellite TV receiving system. But it is really nothing more than a reflector that concentrates the incoming signals to a common focus. The reflector delivers the incoming signal to the focal point where it is collected and passed on to the rest of the system. The signal collection device used to accomplish this is called the "feedhorn."

THE KU-BAND LOW NOISE FEED (LNF)

For optimum performance, the Ku-band feedhorn must precisely complement the performance characteristics of the antenna that it "illuminates." The offset antennas most often used for digital DTH reception come with a feedhorn that creates a "flared" illumination pattern that matches the precise shape of the dish and extends a specific distance or "focal length" from the antenna reflector.

This focal length—the distance between the feed opening or "mouth" and the center of the dish—is a function of the antenna's f/D ratio. In the case of offset antennas, the f/D usually ranges between .6 and .7 for most models.

At the output of the feedhorn is a flange which mates with the waveguide opening of the system's first stage of electronic amplification called the low noise block downconverter or LNB.

Digital DTH systems typically use an all-in-one device called the low noise

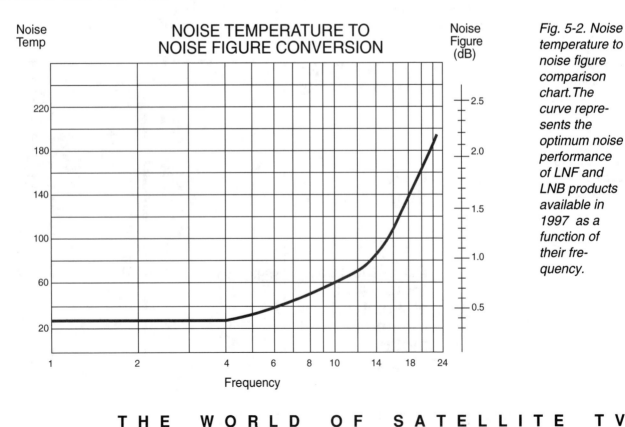

Fig. 5-2. Noise temperature to noise figure comparison chart. The curve represents the optimum noise performance of LNF and LNB products available in 1997 as a function of their frequency.

feed or LNF, which incorporates both the feedhorn and LNB into a single streamlined package. This eliminates the waveguide flange that mates the two components, which is a potential entry point for moisture. Moisture in all its forms, including water vapor, generates a high noise temperature. Its presence inside the feed/LNB assembly will degrade the system performance and may eventually lead to a system breakdown

Within the LNF, the incoming signal funnels down the length of the feedhorn until it reaches a cavity equipped with a short metal probe that is less than an inch long. This is the actual resonant antenna which directly couples the signal to the first stage of electronic amplification.

LNF NOISE CONTRIBUTION

Within the LNF, the first stage of electronic amplification also makes an internal contribution of thermal noise. The internal noise contribution of the amplifier's front-end combines with the thermal noise captured by the antenna and feed. The sum of all noise entering the system sets the system's "noise floor" and is subsequently amplified along with the incoming signal.

Today's Low Noise Feeds use Gallium Arsenide (GaAs) semiconductor and High Electron Mobility Transistor (HEMT) technologies to minimize the internal noise contribution of the LNB, as well as to bring the cost of the product down to affordable levels.

Ku-band LNB noise performance is usually expressed as a "noise figure" that is measured in decibels (dB). The noise figure of an LNF may also be converted to an equivalent noise temperature expressed in degrees Kelvin (K) by using the chart provided. Typical Ku-band LNB noise figure ratings currently range from 0.75 to 1.0 dB, with the lower the noise figure, the better the product performance.

The Ku-band satellites serving the Americas use three distinct Ku-band spectrums: from 10.7 to 11.7 GHz (primarily INTELSAT satellites), from 11.7 to 12.2 GHz (medium-power domestic satellites), and from 12.2 to 12.7 GHz (high-power direct broadcast satellites). All Ku-band digital DTH providers serving the Americas use one of the two latter frequency segments.

The LNF is part of all equipment packages set to receive a single bouquet of digital TV services. It therefore comes tuned to receive those satellite frequencies that the service provider is using. Anyone who is procuring receiving system components on his or her own should take care to ensure that the LNF will match the correct band for the desired satellite or satellites as well as the performance characteristics of the antenna onto which it will be mounted.

In Europe and elsewhere, a wideband LNF can be used to receive signals from all three Ku-band frequency segments. The so-called "universal" LNF can switch electronically between the 10.7 to 11.7 and 11.7 to 12.75 GHz frequency ranges. This wideband LNF must be used with a compatible IRD that can send a switching signal up the coaxial cable to automatically change the incoming LNF frequency range to the spectrum for the satellite TV service that the viewer has selected.

THE C-BAND FEEDHORN AND LNB

The most common type of C-band feedhorn manufactured today is the scalar feed, which consists of two parts: on the outside is a sliding plate with a series of concentric rings, while the inner portion of the feed consists of a fixed open waveguide. The scalar ring

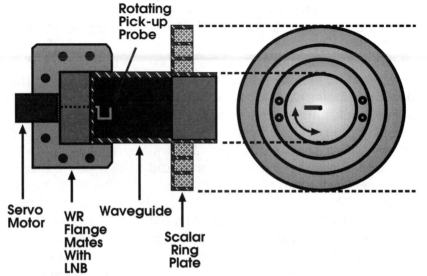

Rotating
Pick-up
Probe

Servo
Motor

WR
Flange
Mates
With
LNB

Waveguide

Scalar
Ring
Plate

Fig. 5-3. C-band feedhorn with scalar ring plate.

design "tapers" the illumination pattern in such a way that, as the feed looks out toward the edge of the dish, it will receive lower signal levels. The rationale for using an "illumination taper" is to prevent the feed from effectively seeing beyond the rim of the dish and intercepting the "hot" noise source of the Earth. With this design approach, most of the signal contribution comes from the inner 70 percent of the dish's surface. Signals coming from the outer 30 percent of the dish are attenuated by 10 to 15 dB. This is opposed to the offset and cassegrain antenna designs where the feedhorn illumination pattern does not require such a steep taper, as the feed points up toward the "cold" sky rather than down toward the "hot" Earth.

The scalar ring plate is usually adjustable so that the installer can set its location along the center waveguide to match feed illumination pattern to correspond with the f/D ratio of the dish. C-band feeds with a fixed scalar ring plate can only be used effectively for a certain range of antenna f/D values. Improper setting of the scalar feed's f/D or use of a fixed scalar feed on an antenna with a f/D outside of the recommended operating range of the feed, will result in either under-illumination or over-illumination of the dish. With under-illumination, the antenna gain, and ultimately the strength of the incoming signal, will be reduced. With over-illumination, the system's noise temperature will increase, thereby re-ducing system G/T, which is the ultimate figure of merit for judging satellite system performance.

Other C-band feedhorn designs also are available in the marketplace. One such design, called the open waveguide approach, does not use a ring plate at all. Another type just has a single fixed plate without any intervening rings. All of these approaches have their adherents and provide comparable performance.

Like their Ku-band counterparts, the C-band feed and accompanying low noise amplifier may be combined into a single unit called the low noise feed (LNF). More often, however, the feedhorn and LNB are sold as two separate units.

C-band LNB noise performance is expressed as a noise temperature in degrees Kelvin (K). The lower the temperature, the less noise introduced into the LNB by its own circuitry. C-band LNBs generally are available in 30 K temperature range, although values down to as low as 14 K are now available.

LNF/LNB GAIN

The specification sheet for any LNF or LNB, regardless of its operating frequency, will also provide the signal gain

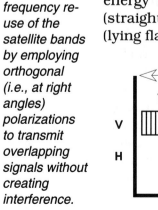

of the product. Measured in dB, the gain rating represents the amount of signal amplification delivered by the product at its IF output port. Just about any consumer-grade LNF or LNB will supply from 50 to 60 dB of gain, multiplying the received signal by one hundred thousand times or more. The object is to supply the IRD with a signal level that falls within the product's range of optimum input values.

An installer might use an LNF with 60 dB of gain whenever the cable length between the LNF and IRD will be longer than what is required for a typical installation. The extra gain will make up for the extra signal attenuation that will take place as the signal passes down the long run of cable. Alternatively, the installer can use an accessory called a line amplifier to boost signal levels by inserting the device in the cable somewhere between the LNF and IRD. The important thing to remember is that the noise rating of the LNF or LNB, either as a noise figure or noise temperature, is the ultimate figure of merit for judging the performance value of the product.

LINEAR POLARIZATION

The signals transmitted by the region's C-band satellites, as well as nearly all the Ku-band satellites operating below 12.2 GHz, are sent by an onboard dish antenna that positions the microwave energy in either a relatively vertical (straight up and down) or horizontal (lying flat) polarization. For best recep-

tion of these signals, the pick-up probe inside the LNF must be precisely oriented in the same plane—horizontal or vertical—as that of the desired transponder. If the orientation of the probe is not exactly matched to the polarization of the satellite transponder, some of the incoming signal will be lost. If the wrong polarization is selected for a particular transponder, you will not receive any signal at all. A slight adjustment of polarization is usually necessary when you switch from satellite to satellite. The amount of variance from one satellite to the next is referred to as "skew."

CIRCULAR POLARIZATION

All high-power DBS satellites, as well as international C-band satellites such as the INTELSAT, Gorizont and Express spacecraft, use an alternate polarization format known as "circular polarization." For the best possible reception of circular polarization, you will need to use a feed or LNF that has been constructed to receive these signals.

Instead of beaming the microwave energy along a "linear" plane, whether vertical or horizontal, circular polarization is transmitted in a helical rotating pattern, with right-hand circular rotating in a clockwise direction as seen from the satellite, and left-hand circular signals rotating in a counter-clockwise direction. Although a standard linear feedhorn can still pick up any signal using circular polarization, half of the available signal power (-3dB) will be

Fig. 5-4. Satellites achieve frequency re-use of the satellite bands by employing orthogonal (i.e., at right angles) polarizations to transmit overlapping signals without creating interference.

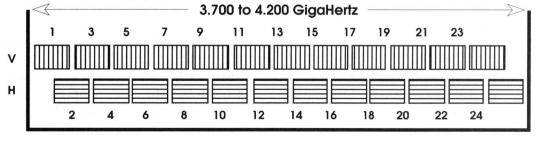

lost. There are several manufacturers that offer a kind of feedhorn that can receive both the linear and circular polarization formats.

POLARIZATION DEVICES

Most commercial satellites maximize their use of the limited frequency spectrums assigned for satellite communications by overlapping the transponders, with their polarization switching from one sense of polarization to the opposite sense every other transponder. This allows twice as many channels in the same amount of space. For example, when a feedhorn or LNF has been adjusted for best reception of a transponder using horizontal polarization, the overlapping transponders using vertical polarization cannot be seen. If the feedhorn inadvertently is set somewhere between vertical and horizontal, more than one TV channel can be seen at the same time, an unsatisfying viewing experience at best.

To select the correct polarization, the LNF or feedhorn may incorporate a small probe that is rotated until best reception is obtained. The probe is rotated by means of a small servomotor that is powered by the IRD. By sensing the strength of the incoming signal, some receivers can select the correct polarization automatically. Many microprocessor-based receivers and IRDs can be programmed to recall the correct polarization format for each individual satellite stored in their memory.

Most digital DTH systems use an LNF that can electronically adjust the polarization of the incoming signal, instantaneously and silently. Rather than adjusting a servo-driven probe, these circuits electronically alter the polarization of the incoming microwave signal.

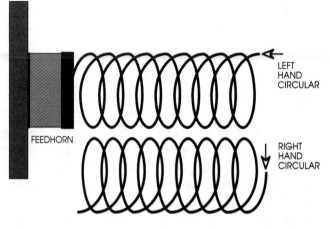

FEEDHORN

LEFT HAND CIRCULAR

RIGHT HAND CIRCULAR

Fig. 5-5. Circular polarization signals.

This introduces a small amount of signal loss, typically 0.1 to 0.2 dB, which for most applications is negligible.

Electronic polarization provides instantaneous polarization adjustment with no moving parts that can give you maintenance problems in the future.

DUAL FEEDS

There are special feeds available which can be used to mount two C- or Ku-band LNBs at the focal point of the dish for simultaneous reception of both senses of polarization. These "orthomode" feeds are used for the installation of systems with multiple indoor units so that each IRD can independently select transponders of either polarization from a given satellite.

HYBRID FEEDS

Dual-band "hybrid" feeds place both the C- and Ku-band feed openings directly at the focal point of the dish so that one antenna can receive satellite TV channels from both C-band and Ku-band satellites. Although the placement of both the C- and Ku-band feed openings at such close proximity to each other results in a slight amount of signal loss at C-band, this loss primarily will be noticeable only when receiving the weaker C-band satellite channels.

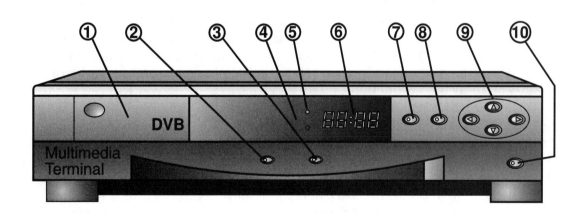

6

THE INTEGRATED RECEIVER/DECODER

The first so-called "integrated" receivers were introduced in the United States back in 1984 when satellite manufacturers began to combine the satellite TV system's receiver, stereo processor, polarizer, and antenna positioner in one streamlined unit. After programmers such as HBO began encrypting their signals back in 1986, a decoder module was added to produce the first integrated receiver/decoder or IRD. This was the first step toward the creation of a consumer-friendly set-top box featuring simple, automatic selection and tuning of satellite TV channels.

Today's IRD (digital, analogue or both) has evolved into a full-blown computer that comes with factory-installed software programs that allow each unit to function as the master control for an entire video home entertainment system. An IRD may now carry so much memory that it is comparable to a desktop personal computer system. The IRD's wireless remote control has become the keyboard and the TV set serves as the computer screen for displaying a variety of flashy, graphic-oriented menus that give the operator total control.

The very complexity of the newest generation of IRDs makes them easier to use than ever before. They are pre-programmed at the factory to keep track of all the settings necessary for receiving

Fig. 6-1. A digital DTH IRD front panel display.

DIGITAL IRD FRONT PANEL

1. Smart-card slot door
2. Information button
3. Guide button
4. Green LED warning light (E-mail, remote control unit, signal reception)
5. Yellow LED warning light (smart card)
6. Four-digit front-panel LED display
7. Menu button
8. OK button
9. Directional arrow button
10. Stand-by button

your favorite satellite TV channels. All you have to do is select your favorite satellite TV channel from a menu displayed on the TV screen and the system does the rest. You don't need to know which satellite transponder is carrying the channel or anything else about how to tune in a program. The IRD remembers and executes all the little adjustments required for perfect reception.

In their quest for simplicity of operation, however, the IRD manufacturers have not forsaken individuals who enjoy having access to a plethora of electronic goodies. If you don't care to keep it simple, there are enough high-tech menus and advanced function commands to keep even the most avid "techie" happy.

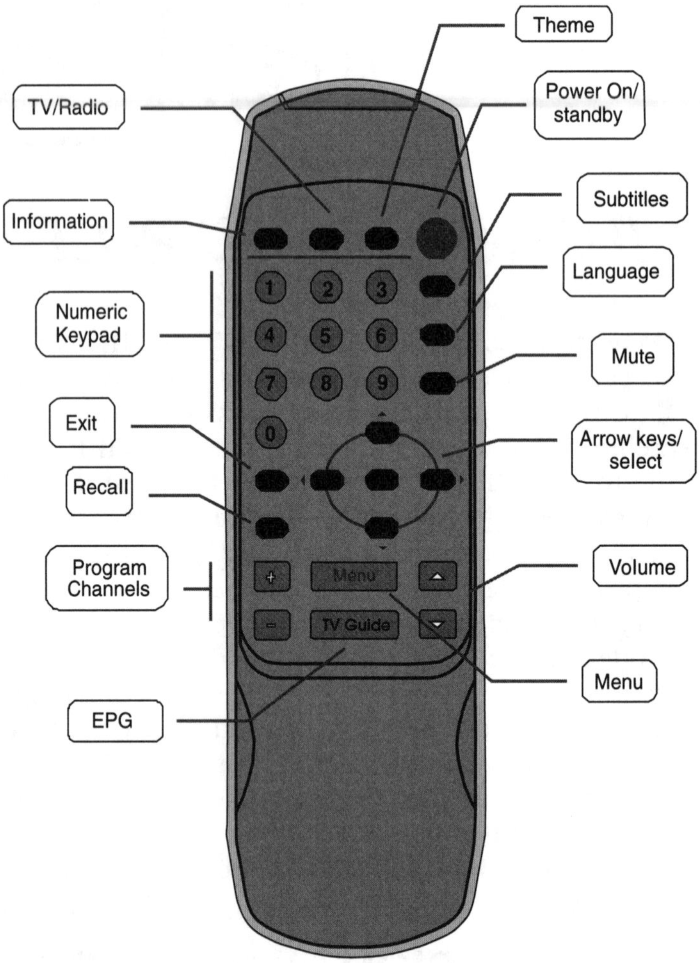

Fig. 6-2.
Remote
control
options.

REMOTE CONTROLS & ON-SCREEN GRAPHICS

Every IRD these days comes with a remote control. A few front-panel controls may still be available, but the hand-held remote commands the vast majority of the available functions. Pushing various buttons on the remote control will bring up on-screen (on the screen of your TV) menus that provide you with a selection of possible choices.

In addition to guiding you through the installation instructions, the IRD displays relevant information on the working status of your system. Change channels, for example, and the name of the TV or audio service will appear momentarily on the TV screen. An on-screen graphics display is a convenient feature to have when viewing satellite program-

ming from a location where you can't see the readout provided on the IRD front panel display. The on-screen graphics and remote control work together to provide you with total hands-free operation of your satellite TV system.

Remote controls come in two garden varieties: infrared and UHF. The infrared remote control transmits an invisible light beam back to the IRD, which has a sensing window on the front control panel. Infrared remotes, however, are limited to line-of-sight operations. There can be no obstacles between the remote and the IRD's sensing window.

UHF remotes use a radio frequency to send the control signals back to an antenna mounted onto the IRD. Since UHF remotes use a radio signal instead of a light beam, the transmissions aren't limited to line-of-sight access. The UHF remote control and on-screen graphics allows for convenient operation of the IRD from different rooms in the house. This is a good option for anyone who has their IRD connected to multiple TV sets throughout the home.

If there is more than one IRD installed in the home, UHF cannot be used to control both units because the remote control for one IRD would interfere with the other. Some hand-held units provide for both infrared and UHF remote control operation. Infrared is used in the room where the main IRD is located, while UHF is used from all other rooms.

REMOTE FUNCTIONS

Hand held remote controls range from the very simple to the very sophisticated. Most IRD remotes allow you access to a wide variety of functions and special features. There can be many different ways of accomplishing the same task, and one IRD may offer features that will not be available from another model. Here is a list of the more common remote functions that you may encounter.

Channel Up/Down. Designated keys on the remote will select which of the available channels you want to see. Individual channels can be accessed in several ways, depending on whether your particular IRD is analog, digital or both. You can push the channel up/down button to surf the available channels or select from a personalized list of favorite channels. In the case of the new digital DTH set-top boxes, press the "TV guide" or "EPG" button to access an electronic program guide, scroll through the list and press the "select" button, and make your program choice.

Volume Up/Down. You can push the volume up/down button on the remote to control the audio level. Many remotes also come with a "mute" button that can be used to zap commercials and other annoying audio phenomena.

Menu. The "menu" button provides the viewer with access to a series of computer-style menus. Menu categories include installation set-up, one or more customized favorite channel lists, an electronic program guide, alternate language audio selection, and optional subtitling. Each menu contains a list of several related functions.

MENU OPTIONS

The following functions commonly are included under one or more IRD menus.

Favorite Video and Audio Channels. Some set-top boxes allow one or more viewers to create customized lists of favorite video and audio services and store these settings in memory. A directory of your favorite video and audio services can be displayed at any time on the screen. Each favorite service is assigned a unique code that the viewer can enter using the remote control. By punching in the number for your favorite service of the moment, the system will automatically select the correct tuning parameters. Alternatively, you can use the up/down keys to scroll through your list and then punch the "select" button to make your choice.

VCR Timer/Automatic Taping. Like VCRs, most set-top boxes now come with circuitry that will permit the taping of multiple events over a period of two or more weeks. These VCR-compatible units have their own internal "real time" clocks. The operator merely selects the channel, "on" time and "off" time for several events, and the IRD will automatically select the channel, and forward the video and audio signals on to your VCR for recording. However, you still must remember to pre-set your VCR to record the selected programs as they occur.

The timer feature also can be used to pre-program the IRD so that small children can view their favorite programs on schedule without needing adult assistance. All the kids will have to do is sit back and enjoy. Full memory protection is provided so that momentary power outages will not preempt the program that you have entered into the IRD's memory.

Parental Lockout and Rating Ceilings. You can designate certain channels off limits so viewers will only be able to access them by keying in a special secret code. This feature can be used to prevent children from gaining access to services that you do not want them to watch. You can also establish a password to control access to certain advanced IRD functions so that the children are unable to wreak havoc on the system when you are out of the room. Some of the digital set-top boxes also allow you to set a motion picture rating ceiling, such as "G" or "PG," so that you don't have to worry what kind of programs the kids are watching when you are not at home.

Surround Sound. This is an audio enhancement that gives spatial depth to the audio broadcasts of major sports events and movies. Surround sound audio processors decode the special surround sound signals and send them to the home entertainment system's front and rear pairs of stereo speakers.

BITING INTO DIGITAL TERMINOLOGY

Coping in the new digital age requires an explanation of a few new terms that have grown up surrounding video compression technology. You might encounter one or more of these terms when evaluating the sales pitches of the various digital DTH services that are out there. If you read and understand the following before any digital shopping expedition, however, you will probably know more about digital TV technology than the salesperson who waits on you. For those who are faint of heart or may suffer from "digiphobia," feel free to skip on to more familiar terrain. As for the best of the brave, get ready to take your first bite!

MPEG-2

In 1988, the International Standards Organization (ISO) established its Moving Pictures Experts Group (MPEG) to create a new standard for the compressed representation of multimedia. This new MPEG-2 standard was approved in 1994.

To efficiently compress video signals, MPEG-2 has several tricks up its sleeve. Each TV scene is comprised of a series of individual images or "frames" of video, with only a small amount of the total picture information actually changing from one frame to the next. Because of this high level of content redundancy, only the parts of any TV picture that change from one video "frame" to the next actually need to be transmitted. What's more, the coding of the picture information can be transmitted in a numerical shorthand form that the IRD can easily reconstruct.

Take this old Beatles song as an example of how statistical redundancy works. "Why don't we do it in the road? Why don't we do it in the road? Why don't we do it in the road? Why don't we do it in the road? No one will be watching us. Why don't we do it in the road?"

For brevity's sake, Paul could have assigned a code for "Why don't we do it in the road?"—such as WDWDIITR—and then coded his message as

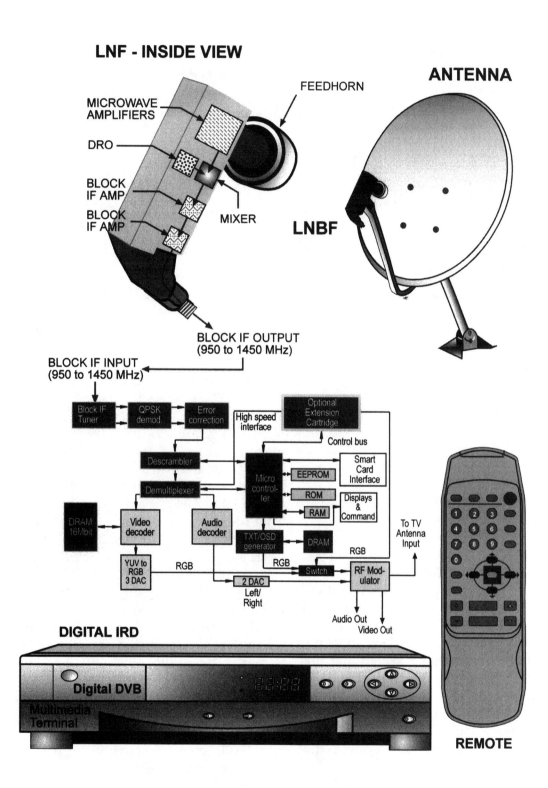

Fig. 6-3. Transformation of the digital DTH signal as it passes through the receiving system.

LNF - INSIDE VIEW

FEEDHORN

ANTENNA

MICROWAVE AMPLIFIERS

DRO

BLOCK IF AMP

BLOCK IF AMP

MIXER

LNBF

BLOCK IF OUTPUT
(950 to 1450 MHz)

BLOCK IF INPUT
(950 to 1450 MHz)

Block IF Tuner

QPSK demod.

Error correction

High speed interface

Optional Extension Cartridge

Control bus

Descrambler

Demultiplexer

Micro controller

EEPROM

ROM

RAM

Smart Card Interface

Displays & Command

DRAM 16Mbit

Video decoder

Audio decoder

TXT/OSD generator

DRAM

To TV Antenna Input

YUV to RGB 3 DAC

RGB

2 DAC
Left/Right

RGB

Switch

RGB

RF Mod-ulator

Audio Out

Video Out

DIGITAL IRD

Digital DVB

Multimedia Terminal

REMOTE

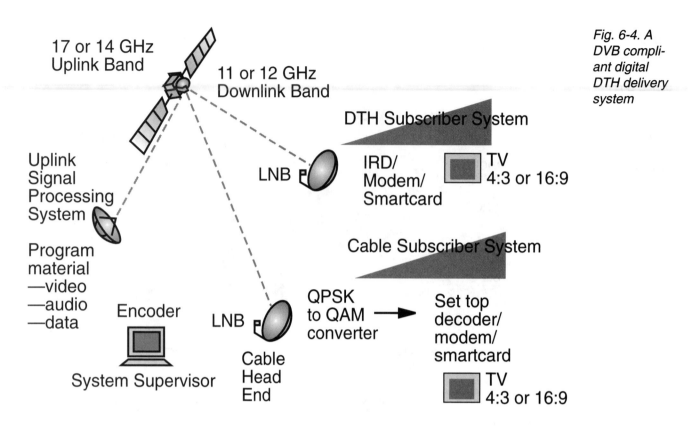

Fig. 6-4. A DVB compliant digital DTH delivery system

"WDWDIITR x 4. No one will be watching us. WDWDIITR."

MPEG-2 also can predict the speed and direction of moving objects so that for some picture changes it merely needs to send the IRD the equivalent of "take the object in square 1 and move it 3 squares to the right and two squares to the top of the screen."

DVB OR NOT DVB:
THAT IS THE QUESTION

Some programmers viewed MPEG-2 as an incomplete standard for broadcasting digital TV signals via satellite. Europe's Digital Video Broadcasting (DVB) Group further clarified just how MPEG-2 should be used for satellite TV broadcasting. Satellite programmers around the world have since adopted its recommendations. Any digital DTH broadcaster or receiving system that fully conforms to the DVB version of MPEG-2 is said to be "DVB compliant."

The term DVB compliant, however, does NOT necessarily indicate signal compatibility between two or more different programmers using the DVB standard. That is because the DVB Group did NOT standardize how the MPEG-2 data stream should be encrypted.

You may hear a salesperson claim that their DVB compliant products offer superior performance over other digital DTH systems that are not DVB compliant. Don't bet the farm on it. The signal quality produced by any digital delivery system is largely a function of the transmission rate assigned to any given video or audio service within the digital bit stream. The quality of what you receive can range from quasi-VHS all the way up to high-definition TV. The final result is up to the broadcaster. From a consumer point of view, the terms "MPEG" and "DVB" mean next to nothing insofar as the issues of signal quality or IRD compatibility are concerned.

The good news about DVB is that it specifies a "Service Information and Teletext" standard that allows each broadcaster to transmit technical stuff that only your IRD needs to know, such as satellite frequencies, channel allocations and other high-tech transmission parameters. Any technical changes to the system are totally transparent to the viewer because the programmer can reconfigure the each IRD's software automatically.

The Service Information and Teletext standard also sets the parameters for the transmission of electronic program guides, which provide a wide variety of information including service provider and channel name; program name, type, and description; alternate channel program lists; and forthcoming program information.

Once installed, the digital IRD will automatically tune to the factory-programmed "default transponder" frequency and access the EPG, Service Information, and conditional access data that it needs before it can begin delivering signals to the TV set. The Service Information data provides the IRD with the picture identification (PID) and sound identification (SID) numbers that the IRD needs to locate every satellite TV and audio service in the bouquet.

DIGITAL BITS AND BIT RATES

Radio and TV signals are composed of electro-magnetic waves of energy, which continuously vary in their levels of frequency and intensity. All of these signals are called "analog" because of the wide signal variance that occurs within any transmission.

Computers, however, convey information in an alternate form consisting of a series of binary ("BI" meaning two) digits or "bits" which correspond to the "on" (1) and "off" (0) states of computer logic circuitry. Any unique string of binary digits can be used to represent whatever we want as long as the receiving system understands the "code words" that we are using to convey the information. Worldwide standards—such as ASCII and HTML (text) or GIFF and JPEG (graphics)—convert information to digital bit strings that are universally understood by all electronic systems receiving the signals. That's why any computer accessing the Internet can view the images presented regardless of whether they are using an IBM, MAC, UNIX, Sun or some other computer platform.

The amount of data information being transmitted in one second of time is called the bit rate, expressed in bits per second (bps):

1,000 (10^3) bps = 1 kilobit/sec (kb/s)
10,000 (10^4) bps = 10 kilobit/sec (10 kb/s)
100,000 (10^5) bps = 100 kb/s
1,000,000 (10^6) bps = 1 Megabit/sec (Mb/s)

SURFING THE DIGITAL MULTIPLEX

Whenever the digital bit stream consists of multiple TV, audio and conditional access data signals, this bit stream is called a "multiplex" by the engineers or a "bouquet" by the programmers themselves. The sharing of common elements, such as the conditional access system, between the various video and audio services is what creates a unified digital "bouquet."

Many of the digital DTH bouquets take advantage of the superior audio transmission quality achieved by MPEG-2, DVB compliant delivery systems to digitally transmit CD-quality audio services as well as TV services. Using a form of digital modulation that is similar to that employed by compact disc players, the new digital broadcasters can deliver unparalleled audio performance directly to your home entertainment system.

The electronic program guide and favorite channel menus will take you to your favorite audio services at the touch of a button on your IRD's remote control.

Within each multiplex, the programmer has considerable flexibility in assigning the bit rates for each individual video, audio and data service. A bit rate of 1.152 Mb/s could be used to transmit a movie as long as a VHS-quality picture was acceptable. A news or general entertainment TV program would require a higher bit rate, such as 3.456 Mb/s. Live sports events are usually sent at a bit rate of 4.608 Mb/s or higher, while a studio-quality broadcast would require a bit rate of 8 Mb/s or more.

The stereo sound channels for each TV service, as well as auxiliary digital audio services, are sent at lower bit rates of 512 kb/s. Also included in the multiplex are the conditional access data that control IRD access to all the available subscription channels.

The digital bit stream is also broken up into a series of individual bundles called "packets," with the content of each packet identified by a special "header" message. The IRD looks at the digital bit stream and only reads those packets that it needs to reconstruct a selected TV or audio channel.

SYMBOLS AND OTHER ESOTERIC STUFF

Now that we've taken our first few bites from a veritable smorgasbord of digital fare, gone surfing on the digital bit stream, and come out the other end still breathing, it's time to hit the hard stuff. No, I am not advocating a visit to your local neighborhood tavern. This is just my way of saying that we are not out of the woods yet.

Digital DTH programmers use a modulation technique called QPSK (short for Quaternary Phase Shift Keying, in case you were wondering). With QPSK, two bits can be processed simultaneously in pairs of 2-bit "symbols." This effectively doubles the data rate over the satellite without any increase in the signaling rate. Digital DTH programmers typically transmit millions of symbols per second. This transmission rate

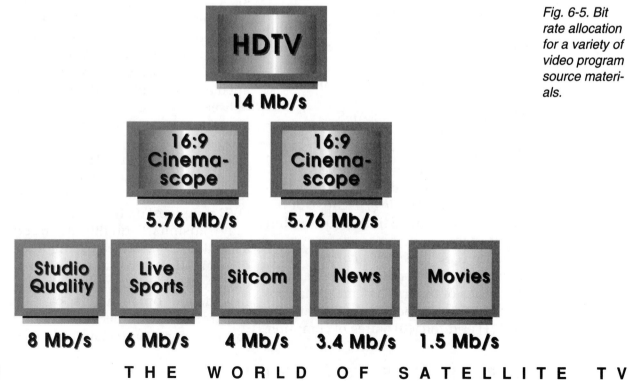

Fig. 6-5. Bit rate allocation for a variety of video program source materials.

is commonly expressed as Megasymbols per second, or Msym/s.

The precise symbol rate in use is an essential piece of information that the IRD must know before it can process the signals. The correct symbol rate for a specific digital bouquet is usually programmed into the IRD's memory at the factory. A subscriber to a single digital DTH service therefore need not know these specific transmission details. Some digital IRDs have the ability to receive multiple bouquets. If you intend to receive more than one digital bouquet with your satellite system, you must select a digital IRD that can receive multiple symbol rates. One multimedia IRD that I know of will automatically select the correct symbol rate for any digital service once it has been tuned to that service's satellite frequency. More often, however, the viewer must manually enter the correct symbol rate into the IRD's memory for each bouquet.

FORWARD ERROR CORRECTION (FEC)

Your digital IRD isn't just satisfied with knowing the correct symbol rate. There is a second piece of information that it also must have before it can begin to process any digital bit stream.

The IRD must have access to special correcting symbols that the encoder adds to the symbols containing the original message. The IRD uses these codes to reconstruct any missing data in the event that all the bits of information sent have not been received. This special coding method is called "forward" (sent ahead to the IRD along with the original message) error correction, or FEC for short.

With FEC, redundant symbols are added to the original message to assist the IRD in reconstructing the message in the event that link noise corrupts some of the symbols being transmitted. Think of FEC much in the same way that you think about the spare tire that you keep in the trunk of your car. That spare

Fig. 6-6. QPSK modulation.

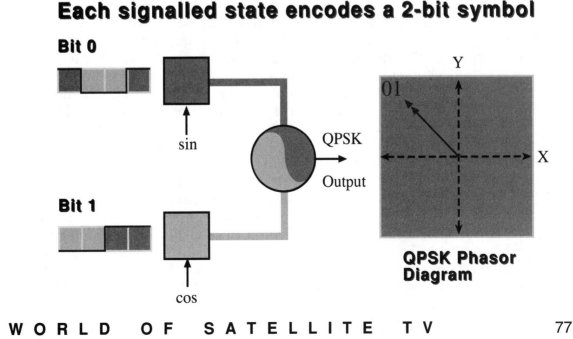

QPSK MODULATION
Each signalled state encodes a 2-bit symbol

Bit 0

sin

QPSK
Output

Bit 1

cos

QPSK Phasor Diagram

tire is only redundant until the time comes when you actually need it, at which point it becomes an essential part for completing your journey.

FEC encoding rates are expressed as a ratio, such as 1/2, 2/3, 3/4, etc. The digit up front indicates the number of original message symbols entering the encoder, with the digit at the back end of the slash representing the number of error-corrected symbols leaving the encoder. For example, the FEC of 3/4 means that for every 3 original message symbols entering the encoder, 4 symbols leave. In other words, there will be one error-correcting symbol out of every 4 symbols being transmitted.

FEC rates usually vary from one digital bit stream to the next. Like the symbol rate, the correct FEC rate for a specific digital bouquet is programmed into the IRD's memory at the factory. For reception of multiple bouquets, the digital IRD must offer some way for the viewer to select the correct FEC rate whenever the receiving system switches from one bouquet to the next. A few digital IRDs will automatically detect the FEC rate in use and reset itself to the new rate.

As we have seen, the digital DTH programmer must use a symbol rate that is high enough to include all of the original message content of the digital DTH multiplex plus the redundant FEC symbols (called the FEC "overhead"). The maximum allowable symbol rate is primarily a function of the available satellite transponder bandwidth. U.S. DBS operators typically use symbol rates of about 20 Msym/s because the available transponder bandwidth from a DBS satellite is only 24 MHz. Higher symbol rates of 30 Msym/s or more can be used over a wideband (54 MHz) transponder.

Each digital DTH programmer makes certain trade-off decisions when setting up its service. How many TV services will be placed on each transponder and what will be the bit rates assigned to each service? The choice is a balance between the smart economics of squeezing as many transponders as possible onto each satellite transponder versus the good engineering of limiting the number of services per transponder so that the video quality of each service is heightened.

The programmer can select an FEC rate with a low level of redundancy, such as 7/8, and accept a decrease in its signal availability due to rain fades. Or it can opt for a high level of redundancy by selecting an FEC rate with a high level of redundancy, such as 1/2, so that the system will perform well in heavy rain rate areas. It is the flexibility that both the MPEG-2 and DVB standards give to programmers that accounts for the wide variance in symbol and FEC rates that satellite TV viewers encounter around the world today.

DIGITAL IRD THRESHOLD

The digital IRD does not have any noise reduction or TI filters built into it. With an analog satellite TV signal it is possible to reduce the impulse noise or "sparklies" present in the video coming from a low-power C-band satellite by narrowing the pass band of the IRD. An analog IRD may also come with built-in notch filters for eliminating C-band terrestrial interference. Digital DTH signals, however, cannot be filtered at all without losing essential data components.

When you are receiving a digital DTH service, you either have a perfect picture—that is, one without any impulse noise or "sparklies"—or you have no picture at all. In this case, it is like a light switch: either on or off. The point at which the digital IRD switches between

Fig. 6-7. The
threshold of a
digital IRD is
like a light
switch: either
on or off.

DIGITAL IRD THRESHOLD

The Threshold for a digital satellite receiver is defined as occurring at a particular bit error rate (BER). Reduction in incoming signal level (Eb/No, expressed in dB) has no effect on video S/NR. Instead, video signal quality is determined by the bit rate and frame resolution assigned to each individual TV service within the MPEG-2 digital multiplex. Blocking and freeze frame artifacts, however, will occur as a rain faded Ku-band satellite signal approaches threshold, just before the digital receiver switches off.

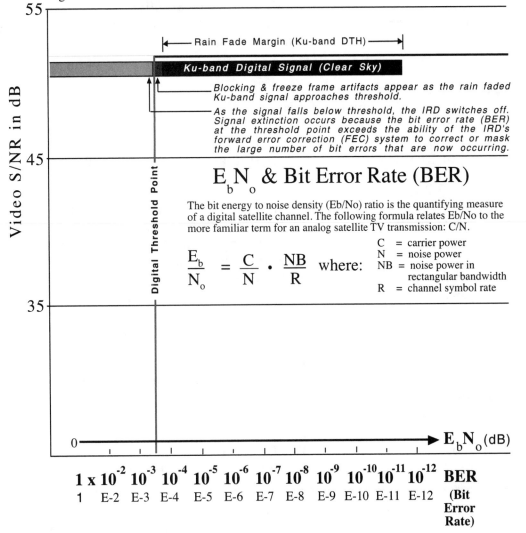

the on and off states is called the digital threshold. IRD manufacturers usually define digital threshold as a "bit error rate" (BER) such as 5E-4 (5.0 x 10^{-4}).

Measured in exponential notation, the Bit Error Rate (BER) quantifies the performance level of the digital link. A BER of 1×10^{-3} expresses the probability of one bit error occurring in a block of 1,000 bits. A BER of 5.0 x 10^{-5} is superior to a lower BER of 9.0 x 10^{-4} because there is a probability that less bit errors will occur within any given time frame. BER also may be expressed in an alternate form: 5E-4 or 3E-3, or the equivalent of a BER of 5 x 10^{-4} or 3 x 10^{-3}.

REPRESENTATIVE DIGITAL IRD RECEIVER SPECIFICATIONS

TUNER

Input Frequency:	950 to 1450 MHz
Input Impedance:	75 ohms
1st IF:	479.5 MHz
2nd IF:	70 MHz
IF Bandwidth:	24 MHz (DBS) or 36 MHz
Polarity Switching:	13/17 volts. *Allows IRD to switch LNF polarization.*

VIDEO PERFORMANCE

Video frequency range:	5 Hz to 5 MHz
Output Impedance:	75 ohms
Output Level:	1 volt peak-to-peak
Output Display:	4:3 or 16:9 aspect ratio, selectable

DEMODULATOR

Type:	QPSK (Quadernary Phase Shift Keying)
Input Data Rate:	15 Msym/s to 30 Msym/s *(A symbol rate lower than 15 Msym/s may be required to receive some digital services.)*
Informational Rate:	49 Mbits/second for six video channels
Forward Error Correction:	Inner code FEC - convolutional Outer Code FEC - Reed Solomon *(the MPEG-2 standards for FEC)*

AUDIO OUTPUTS & AUDIO DECODER

Audio Output Level:	1 volt peak-to-peak into 75 ohms
Frequency Response:	20 Hertz to 20 kHz
Dynamic Range:	75 dB
Sampling Rate:	32, 44.1 and 48 kHz
Audio Syntax:	MPEG 2 DVB
Converter:	18 bit precision Delta-Sigma

ANALOG IRD FEATURES AND OPTIONS

In addition to the standard features described earlier in this chapter, many analog IRDs come with a dazzling array of "bells and whistles," exotic features that can enhance the performance of your satellite TV system. The following are some of the more commonly encountered optional features.

Auto Install. The installer merely has to find the first two satellites and the IRD will automatically locate the remaining satellites and program their locations into memory.

Auto Peak. After each satellite has been located, the IRD automatically selects the correct polarization format, skew and video center frequency for every channel. The operator also can select auto peak at any time to readjust the antenna, polarization, skew, and channel settings for best performance.

Unfortunately, the auto peak feature is not perfect. When using auto peak, observe the reading of your IRD's signal strength meter to insure that maximum system performance has been obtained. This meter either will be found on the IRD front panel or as part of the on-screen graphics menu. You can always use the IRD manual controls to touch up any adjustments made after initiating an auto peak, if necessary.

TI Filters. Many set-top boxes also come with built-in filters that can be used to eliminate interference generated by terrestrial microwave links that share frequencies with the C-band satellites. These terrestrial interference (TI) filters can be engaged manually by the operator whenever needed. Filter settings also can be stored in memory on a satellite-by-satellite and channel-by-channel basis.

Threshold Extension and Video Noise Reduction Filters. These filters can be used to improve IRD performance when viewing some of the weaker analog TV services. Threshold extension filters also can enhance the quality of reception from certain Ku-band satellite TV services that are transmitted within a much narrower frequency bandwidth (24 or 27 versus 30 or even 36 MHz) than their C-band counterparts.

Analog IRD manufacturers try to maximize the incoming signal strength while keeping both external and internal noise to a minimum. This relationship is expressed as the Carrier to Noise Ratio (C/N or CNR).

Every IRD has a threshold point (expressed in dB C/N). When the C/N falls below this point, the video rapidly becomes noisy. Once the C/N is a dB or more above threshold, impulse noise or "sparklies" disappear. The lower the IRD's threshold rating, the better. It is usually about 7 to 8 dB C/N.

These filters lower the C/N of the IRD by using specially shaped bandpass filters and other processing circuitry. Lowering the IRD's C/N threshold can have the same effect as if you used a lower noise temperature LNB or bought a larger antenna.

Satellite and Channel Formats. Most analog set-top boxes available these days have been designed to effectively process both C- and Ku-band signals. The IRD, for example, automatically chooses a 24-channel format whenever a C-band satellite has been selected or a satellite specific Ku-band transponder scheme if a Ku-band satellite has been selected.

Fig. 6-8. The threshold of an analog IRD is the point at which the relationship between satellite signal C/N and video signal S/N becomes non-linear.

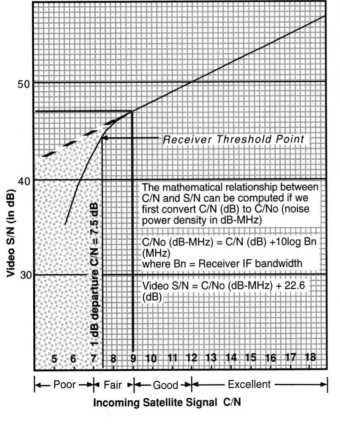

The mathematical relationship between C/N and S/N can be computed if we first convert C/N (dB) to C/No (noise power density in dB-MHz)

$$C/No \text{ (dB-MHz)} = C/N \text{ (dB)} + 10\log Bn \text{ (MHz)}$$
where Bn = Receiver IF bandwidth

$$Video \text{ S/N} = C/No \text{ (dB-MHz)} + 22.6 \text{ (dB)}$$

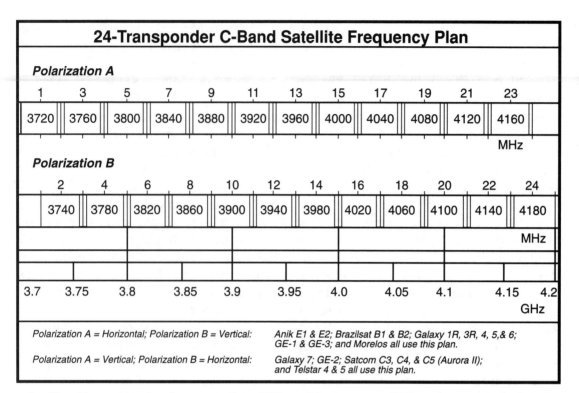

Fig. 6-9. The standard 24-transponder plan used by most C-band satellites serving the Americas.

Audio Format. Analog satellite TV signals usually carry the audio portion of the TV program on one or more subcarriers that are offset from the frequency of the main video carrier. Additional subcarriers on any transponder can also carry radio programs or other "auxiliary services." The audio menu is selected when tuning in to the desired subcarrier on any satellite TV channel. The subcarrier frequencies most commonly used by American C-band satellites are 6.2 and 6.8 MHz, while international satellites may be using 5.8, 6.6, 6.65 or 7.0 MHz subcarriers to carry the main TV sound signals. Auxiliary services may be found anywhere between 5.0 and 8.5 MHz.

For full stereo sound to be produced by your IRD, the correct stereo mode and stereo subcarrier frequency pairs must be selected and programmed into memory for each available service. Some IRDs may have had the stereo audio settings for existing services programmed into memory at the factory.

Most stereo IRDs also provide for the selection of a wide (400 kHz), medium (200 kHz) or narrow (120 kHz) audio filter. While the monaural audio for any satellite TV channel also may be carried in the wide audio format, stereo TV broadcasts and auxiliary subcarrier services often require a narrow audio format. Engaging the narrow band audio filter will enhance the reception of these narrow band audio services.

Most IRDs come from the factory with the correct audio bandwidth and pre-emphasis settings stored in memory for each channel. However, you may need to make your own adjustments for new services or whenever existing services move to a new transponder or satellite.

INTERNATIONAL IRD MENU OPTIONS

A few analog IRDs have been designed specifically for receiving TV signals from the available international satellites, which often transmit signals which employ video and audio formats that differ from what is typically encountered on the North American domestic satellites. There are a few international IRDs that will come preprogrammed from the factory with all of this information stored in memory. Even so, anyone who intends to receive these international satellite transmissions should be prepared to make manual adjustments to the IRD to set the correct video and audio parameters on a channel-by-channel basis.

Tuner selections. The international satellites do not conform to the standard 24-transponder plan shown on the left. In this case, the channel frequency for each satellite service can be fine tuned through the use of an on-screen menu. Picture fidelity also can be optimized by adjusting the IRD's IF bandwidth and video noise reduction controls. Fine tuning of the polarization "skew" setting may further optimize the reception.

Video selections. The receiver must be programmed to receive the correct video standard used by each satellite TV channel. A multi-standard TV set will also be required to view those satellite TV services that use a TV standard that differs from the NTSC standard used in North America. What's more, the video level should be set to either the LOW or HIGH setting. Additional details on world

TV standards and the equipment required for their recpetion is provided later in this book.

Audio selections. Sound fidelity is enhanced by selecting the correct audio de-emphasis for each channel (50, 75 and 125 µs or the European J17 standard). The correct audio bandwidth for each international satellite TV service also must be programmed into memory: either 125, 230 or 400 kHz. The operator also will choose the correct audio format (monaural or stereo) and the correct audio subcarrier frequency or stereo frequency pair for each available TV service.

THE BEST OF BOTH WORLDS

A new-generation IRD is now available that will provide satellite TV viewers in North America with access to both analog AND digital satellite TV services. Called the 4DTV, this new IRD from General Instrument is capable of tuning in a 600-channel universe when installed on a large-aperture steerable C-band dish. The unit includes the analog decoder required for reception of more than 150 C-band pay TV services using the VideoCipher-RS encryption system plus a digital decoder for reception of 100 other digitally-compressed C-band pay TV services using the DigiCipher encryption system. The 4DTV also can receive 75 free TV services on C- and Ku-band satellites serving the U.S., Canada and Mexico, as well as 100 free CD-quality digital audio channels, and more than 100 free analog audio services that are broadcast on audio subcarriers.

ANALOG IRD SPECIFICATIONS

For those consumers not familiar with electronics, the technical specifications advertised within the sales literature can be baffling. Here is a list of some of the common analog IRD specs and what you should look for.

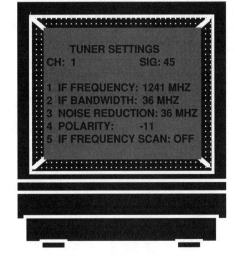

Fig. 6-10. Typical tuner setting on an international IRD.

TUNER SETTINGS
CH: 1   SIG: 45

1 IF FREQUENCY: 1241 MHZ
2 IF BANDWIDTH: 36 MHZ
3 NOISE REDUCTION: 36 MHZ
4 POLARITY: -11
5 IF FREQUENCY SCAN: OFF

ANALOG SATELLITE TV IRD SPECIFICATIONS

RF

Input Frequency:	950 to 1450 MHz
Input Impedance:	75 ohms
2nd IF:	70 MHz
IF Bandwidth:	27 MHz
Threshold:	Less than 7 dB C/N

VIDEO PERFORMANCE

Standard:	NTSC
No. of Channels:	24 at C band; up to 32 at Ku band
Output Impedance:	75 ohms
Output Level:	1 volt peak-to-peak

AUDIO PERFORMANCE

Audio IF Bandwidth:	130/230/400 kHz selectable
Subcarrier Frequency:	5.0 to 8.5 MHz
Audio Modes:	Mono & Stereo
Preemphasis:	50, 75, 125 µs or J17 selectable
Peak-to-Peak Deviation:	100 kHz
Frequency Response:	+0/-3 dB 50 Hertz to 15 kHz

RF MODULATOR

Output Frequency:	VHF (Ch. 3/4) or UHF (Ch. 32~40)

POWER REQUIREMENTS

Primary Power:	110/120 volts a.c., 60 Hz (in North America)
	220/240 volts a.c., 60 Hz (other countries)
Power to LNB:	18 Volts d.c. @ 300 mA Nominal
Polarizer Output:	5 Volts d.c. @ 500 mA Nominal

ACTUATOR INTERFACE

Motor Power:	36 Volts d.c. nominal, 3 Amp. maximum
Sensor Type:	Pulse, compatible with reed, relay, optical disk or hall-effect switch; 5 Volts d.c. available for optical disk and hall-effect sensors.

7

THE EVOLUTION OF BROADCAST TV TECHNOLOGY

All of today's pay TV networks use some form of encryption to secure the video and audio components of their program services to maintain control over the distribution of their signals. Subscribers to these networks must use some form of decoder to reconstruct the original picture and/or sound information. Before we can contemplate the many different aspects of video encryption, however, we must first understand some of the basic characteristics of a standard analog TV signal. The following chapter provides an overview of the evolution of the broadcast TV signal, from the inception of broadcast television itself to the more recent introduction of encrypted pay TV services and the long anticipated arrival of a new digital TV (DTV) standard for the United States.

THE FIRST TV TRANSMISSION SYSTEMS

Philo T. Farnsworth is generally credited as the first experimenter to transmit TV signals electronically. In 1928, he relayed the world's first electronic TV signals to astounded viewers at the San Francisco Merchant Exchange from his laboratory on Green Street, which was more than a mile away.

The early TV cameras developed by RCA's Vladimir Zworykin converted moving images into hundreds of separate horizontal lines, with the more lines per complete image or "frame," the greater the resolution or clarity of the TV picture. These early experimental TV transmissions delivered monochrome (black and white) images that used either 343 or 441 lines to produce each full frame of video. The low-resolution images that these systems produced, however, clearly were not suitable for commercial broadcast applications.

On July 31, 1940, the U.S. Federal Communications Commission (FCC) convened a group of 168 electronic specialists and charged them with the task of setting a common video transmission standard for TV broadcasting in America. The National Television Standards Committee (NTSC) responded by submitting a series of recommendations that delineated a broadcast TV signal consisting of 525 scanning lines per video frame transmitted at a rate of 30 frames per second. Moreover, each frame would consist of two alternating "fields" consisting of 262.5 lines each. The NTSC concluded that the speed at which the two "interlaced" fields would alternate—60 times per second—was so fast that the human eye would perceive the two alternating fields as a single image. A 4:3 picture "aspect ratio," the ratio of screen width to screen height, was also selected. On March 8, 1941, the FCC approved these recommendations, thereby establishing what is now known as the NTSC video standard.

THE MOVE TO COLOR TV

The transmission of color TV signals did not become technically possible until the late 1940s. CBS initially developed a 405-line color TV system that used a frame rate of 24 frames per second, the same frame rate used by motion pictures. However, the CBS system was totally incompatible with the era's black-and-white TV sets, which complied with the 525 scanning lines and 30 frames per second recommendations of the NTSC.

The FCC decided to reconvene the National Television Standards Committee to recommend the technical parameters of a color TV transmission system that would be backwards compatible with the existing monochrome TV sets then in use. The solution was to use a secondary "subcarrier" frequency embedded within the monochrome TV signal to carry the color or "chrominance" component of the TV signal. Older monochrome TV sets would therefore be able to tune in to color TV broadcasts and receive a black-and-white picture.

The NTSC also decided to allow the amplitude of the chroma-color subcarrier to determine the image's color saturation level, while the color's "tint" would be determined by comparing the chroma-color subcarrier's "phase" to a reference signal generated within each color TV set. In late-1953, the Commission subsequently ratified the NTSC's recommendations, establishing a new

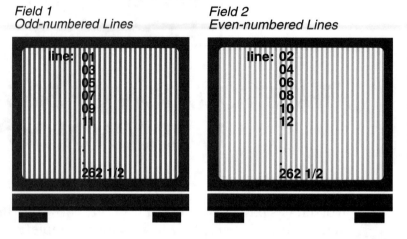

Each frame of video is composed of two fields that alternate at a rate of 60 times per second or 60 Hz

Field 1
Odd-numbered Lines

line: 01
03
05
07
09
11
.
.
262 1/2

Field 2
Even-numbered Lines

line: 02
04
06
08
10
12
.
.
262 1/2

Fig. 7-1. The 525-line NTSC standard uses interlaced video, where one complete image or "frame" is composed of two alternating fields of 262.5 lines each.

U.S. color TV transmission standard. Most of the other countries in the Western Hemisphere have since adopted the NTSC system, as well as Burma, Japan, the Philippines and Taiwan.

It took twelve years for the U.S. household penetration of color TV sets to reach 4.9 percent. In 1965, however, the three major TV broadcast networks began offering their affiliate stations a substantial amount of color TV programs during the so-called "prime time" viewing hours. This fueled the demand for color TV sets so that by 1970, color set penetration had jumped from 4.9 percent to 35.4 percent, and six years later it reached 73.6 percent.

NEVER TWICE THE SAME COLOR

Not all viewers were ecstatic about the quality of the NTSC color system. Back when I was a teenager, I recall walking through a TV store that had a bank of TV sets all tuned to the same TV program. I stopped to take a look and immediately noticed wild color variations from one TV set to the next. The reference signal used to determine the shade of color varied from one TV set to the next, which is why all NTSC color TV sets have a "tint" control that the viewer

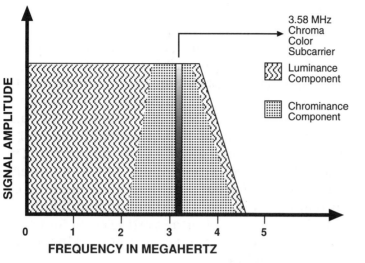

Fig. 7-2. 525-line NTSC video baseband spectra showing the luminance (brightness) and chrominance (color) components.

NTSC video signal.

During the 1950s, European engineers also complained about the low resolution of the NTSC standard's 525-line images, which is why Europe adopted two alternate higher-resolution standards dubbed PAL (Phase Alteration by Line) and SECAM (Sequence with Memory). Each of these systems features 625 lines per video frame with a frame rate of 25 frames per second. Alternate techniques for transmitting the color component of the video signal also were adopted. All of Europe and the Middle East—as well as most countries in Africa and Asia—have officially adopted either the PAL or SECAM video standard.

Nevertheless, the PAL and SECAM standards have their own unique problems. This gave the Americans the face-saving opportunity to comment that PAL actually stood for "pretty awful looking" and SECAM was just the normal French way of producing a "system essentially contrary to the American method."

The introduction of three different analog video standards has given broadcasters major technical difficulties over the years. These difficulties first became a major concern when satellites began to make it possible for TV signals to be transmitted live from one part of the world to any other. The various national broadcast networks worldwide were impelled to develop special standard converters and "transcoders" to convert live or taped foreign TV broadcasts to the local video standard.

can manually adjust. To this day, TV engineers from around the world refer derisively to NTSC as actually standing for "never twice the same color."

NTSC also suffers from unwanted interference byproducts called "picture artifacts" that reduce the quality of the TV image. These artifacts are caused by the interaction between the luminance and chrominance components of the

Fig. 7-3. 525-line NTSC video test signal displayed on a video waveform monitor showing the active picture interval, horizontal sync pulse and color burst.

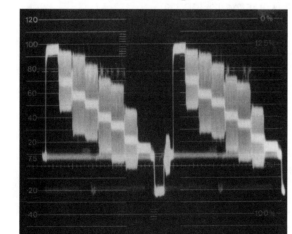

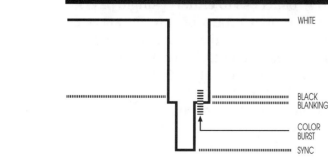

THE MOVE TOWARD ENCRYPTION

It wasn't until the advent of the "Satellite Age" that broadcasters were forced to make major changes in the way that they transmitted their TV signals. On November 5, 1982, Canadian Satellite Communications Inc. (CANCOM) became the first full-time TV programmer to encrypt four of their TV services relayed by Canada's Anik D1 satellite. By the end of 1985, more than one million satellite TV systems were installed and operating across North America. Several other programmers who were using satellites for cable TV distribution purposes began to consider how they might prevent the growing C-band satellite dish market from accessing their pay-TV signals.

On January 15, 1986, Home Box Office became the first U.S. pay TV operator to use encryption when it began scrambling its HBO and Cinemax feeds to cable TV affiliates nationwide. Other programmers, such as CNN Headline News, Showtime and the Movie Channel, also followed suit. Due to legislation passed by the U.S. Congress in 1984, however, all pay TV programmers using encryption technology were obligated to offer their services to any home dish owner who was willing to pay for authorized access.

THE TECHNOLOGY OF ENCRYPTION

All broadcast TV encryption systems have three primary components: an encoder, an authorization center, and a universe of low-cost, individually addressable decoders.

The encoder, which is located at the satellite uplink facility, converts the analog TV signal into an encrypted message that must be decoded before it can be displayed on a standard TV set. The encoder is also linked to a computerized authorization center that can process subscription orders and control all decoders in the system.

Total control is achieved by assigning a unique address code to each subscriber's integrated receiver/decoder or IRD. The authorization center uses this code to turn on, or turn off, any individual IRD, or even to control a selected group of set-top boxes. The authorization center, for example, can "black out" specific regions of the country when a particular programmer is transmitting a sporting event for which it does not own the national broadcast rights.

Whenever I refer to a set-top box as an analog IRD, I am actually referring to the type of satellite signal that the IRD is capable of receiving. As we shall see below, the so-called analog IRD must be capable of processing a variety of digital instructions before it can deliver any analog-based satellite TV program to the viewer's TV set.

VIDEOCIPHER RS

In North America, the VideoCipher-RS (for Renewable Security) encryption system developed by General Instrument is widely used by the satellite broadcasting community to transmit pay TV subscription services to receiving dishes equipped with analog-style, set-top boxes. The VideoCipher RS encryption system, or "VCRS" as I shall call it hereinafter for brevity, can accommodate up to 256 "tiers" of programming. Each tier can be used to control access to either a single pay-TV service or a group of pay-TV services that are marketed as a unified program package. Moreover, VCRS has the capability of individually addressing up to 50 million set-top boxes. VCRS also can employ electronic countermeasures, whenever required, to disconnect any set-top boxes that have been illegally

Fig. 7-4. 525-line NTSC horizontal and vertical blanking intervals.

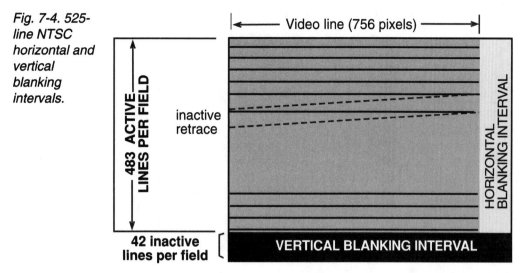

483 ACTIVE LINES PER FIELD

inactive retrace

Video line (756 pixels)

HORIZONTAL BLANKING INTERVAL

42 inactive lines per field

VERTICAL BLANKING INTERVAL

modified to receive encrypted TV programs for free.

VCRS also features an electronic "smart card" that provides the essential conditional access (CA) data that the IRD needs before it can decode the encrypted signals. This smart card, which plugs into a CA card reader slot located on the front panel of the IRD, contains a single Large Scale Integrated Circuit (LSIC) chip that holds the mathematical algorithms or "keys" that unlock each IRD in the system.

Encrypting the Video Signal. All TV sets, whether NTSC, PAL or SECAM, contain a cathode ray tube (CRT) that uses an electron gun to shoot a stream of electrons at the interior wall of the TV screen. This beam of energy excites the screen's phosphor coating, which produces the images that we see.

When the electron gun reaches the end of each video line, it must be told when to switch off so that it can invisibly sweep back across the screen and begin to trace the next video line. A sync pulse located in the TV signal's "horizontal blanking interval" (the horizontal timebase) provides this information. Another sync pulse located within a corresponding "vertical blanking interval" (the vertical timebase) instructs the

electron gun to switch off and return to the top of the screen.

The VCRS encoder removes both of these sync pulses, with the result that the TV set has no way of determining precisely when to end one video line or field and begin the next. The encoder also inverts the TV signal's luminance component to a negative state and also shifts the location of the chrominance signal component, further confusing the TV set.

To make amends, the encoder digitally delivers a set of instructions that tells the IRD exactly when to insert the missing sync pulses, where to find the missing color information, and just how to return the luminance component of the video to the normal "upright" state. However, this digital set of instructions is encrypted so that the signal can only be resolved by those set-top boxes that have the electronic keys required to correctly read these instructions.

The General Instrument factory electronically programs a unique twelve-digit unit address into each IRD that it manufactures. The TV screen will display each IRD's address code whenever the viewer uses the remote control to enter the "SETUP 1" command. Each programmer requires this address number whenever a new subscriber wants to order an encrypted TV service.

The programmer provides the IRD's address code to the General Instruments DBS Authorization Center in La Jolla, California. The Authorization Center in turn will send an authorization message over the satellite that provides the IRD with the set of instructions that it needs to begin decoding the service. Alternatively, the Authorization Center can remove the authorization message from the data stream to shut off any individual IRD. Either process can be completed within a matter of seconds.

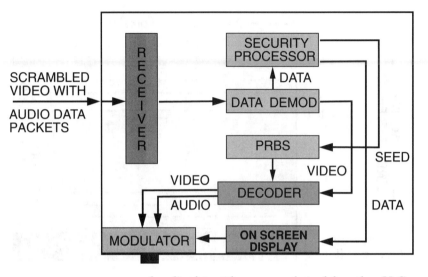

Fig. 7-5. Block diagram of an intergrated receiver/decoder (IRD).

VCRS also includes a "SETUP 0" command that will display diagnostic data on the TV screen. In the event that subscribers are unable to receive a TV program service for which they have already been authorized, the Authorization Center may ask the subscriber to provide some of these codes to assist in determining the cause of the service interruption.

Encrypting the TV Audio. The VCRS encoder is not merely satisfied with transforming the video component of the TV signal into a jumbled mess. It also digitizes the audio and then inserts the resulting digital bit stream into the encrypted TV signal's horizontal blanking interval (HBI), which has enough space to accommodate one stereo audio channel plus a utility data channel.

The encoder also encrypts the bit stream by adding each digitized audio sample to a sequence of binary numbers that have been generated by the encoder's pseudo-random binary sequence generator, or PRBS. This random binary sequence is the "electronic key" that the IRD needs to restore the audio to its original state. In the case of all U.S. encryption systems, this is the 56-bit DES (for Digital Encryption Stan-

dard) algorithm mandated by the U.S. National Bureau of Standards.

To generate this 56-bit key, the IRD must also contain a PRBS that can be synchronized with the PRBS at the encoder so that both units are producing identical strings of random numbers. The encoder transmits information over the satellite, called the "seed," that provides each IRD's PRBS with the required synchronization information.

The encoder also encrypts the seed so that the IRD must also have an electronic key for unlocking this synchronization information. The smart card is the system component that supplies this function.

Similar to the digital set-top boxes discussed in the previous chapter, VCRS also transmits forward error correction (FEC) data to assist the IRD in detecting and correcting errors caused by link noise. The FEC data is transmitted within the HBI along with the audio's digital bit stream. VCRS-compatible, set-top boxes can detect and correct single-bit errors as well as conceal or "mask" double-bit errors.

OTHER ANALOG ENCRYPTION SYSTEMS
There are several other encryption systems that satellite dish owners in

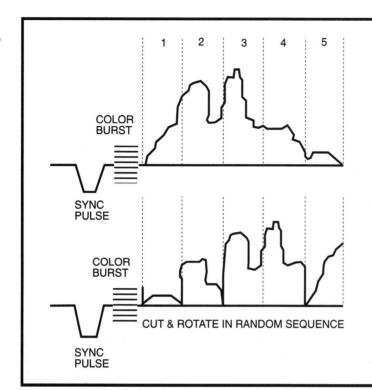

the Americas may encounter as they travel along the satellite superhighways. Let's take a brief look at a few of the more commonly encountered ones.

VideoCipher I. The "VCI" system is exclusively used to encrypt satellite TV signals that are not intended for viewing by the general public, such as news and sports "back-hauls" from the site of a live event to a TV network's home studios. There is no such thing as a consumer-grade IRD for VCI.

VCI is an improvement over VCRS in that it digitally encrypts both the video and audio components of any satellite TV broadcast. VCI uses what is known as the "cut and rotate" method for encrypting the video signal. The encoder samples each video line and converts each line segment to an equivalent digital message. The encoder also determines the cut points for each digitized line and then rotates each line segment so that it can be inserted between alternate cut points. The location of the cut

points varies from line to line and all vertically oriented picture information is stepped back and forth across the screen in a sequence that changes from one field to the next. The audio component is encrypted much in the same way that VCRS encrypts the audio for the services that it handles.

Like VCRS as well as most other modern encryption systems, VCI uses synchronized pseudo-random binary sequence generators at the uplink and downlink locations to generate the algorithms required for encrypting the video and audio signals. With VCI, the IRD receives the set of instructions needed to perform complementary cut-and-rotate operations that reassemble the image into its original state. The VCI encoder uses the vertical blanking interval (VBI) to send instructions to the IRD concerning the location of the cut points.

B-MAC. Developed by Great Britain's NTL and exclusively licensed to Scientific Atlanta, B-MAC (for Multiplexed Analogue Component, Type B) is used by satellite service providers worldwide to encrypt their TV program services. At one time or another, B-MAC has been used for DTH operations in several countries including Australia, Indonesia, South Africa and the United States. Within North America, however, B-MAC hasn't been used for DTH applications since Primestar switched to an all-digital delivery system in the mid-1990s.

B-MAC uses an encryption technique known as the "line translation" method, where each video line is delayed in a random fashion by several microseconds. This creates the familiar cross-hatched, diamond-shaped patterns that satellite dish owners the world over encounter occasionally.

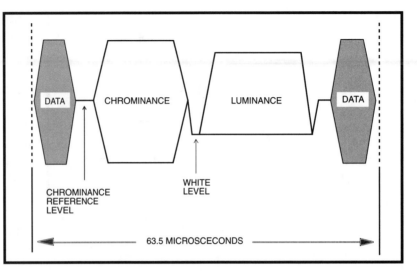

Fig. 7-7. The sequential transmission of the chrominance, luminance and data information is one of the key features of B-MAC.

The B-MAC encoder uses the VBI to send addressable packets to its IRD universe, while the HBI is used to transmit as many as six digitally-encrypted audio channels plus one utility data channel.

In addition to its role as an encryption system, B-MAC also eliminates some of the inherent flaws with NTSC that were mentioned earlier in this chapter. For example, B-MAC totally eliminates the picture artifacts caused by the "cross-talk" between the luminance and chrominance components of the NTSC signal. With B-MAC, the chrominance information is transmitted within one-third of the active scanning time, with the luminance information contained in the remaining two-thirds.

Viewguard. This is yet another encryption system that is not used for DTH applications. Broadcast networks around the world use this system to encrypt their news and sports back-haul feeds. Like VCI, the Viewguard system digitally encrypts video through the use of the cut and rotate method. The audio component of the satellite TV signal also is digitally encrypted.

DIGITAL DTH ENCRYPTION

The available encryption techniques increase dramatically whenever an analog TV signal is transported into the digital domain. As I pointed out in the last chapter, DVB-compliant digital DTH systems can be rendered unavailable by restricting access to compatible set-top boxes as well as any information concerning the symbol and FEC rates in use. Digital DTH signals also are trans-

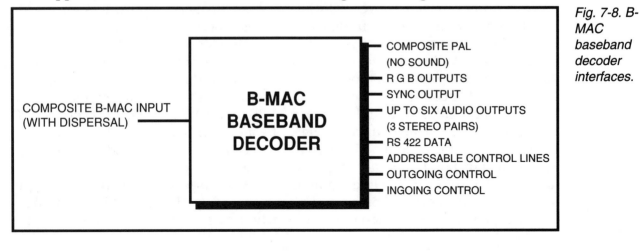

Fig. 7-8. B-MAC baseband decoder interfaces.

mitted in addressable packets that can be subjected to algorithms in a greater number of ways than what is possible when working in an analog domain. For example, the conditional access data for a digital DTH transmission does not have the bandwidth constraints imposed on analog encryption systems, which must insert the CA data into the analog TV signal's HBI and/or VBI.

With analog encryption, the satellite TV viewer will, at the very least, see a jumble of squiggles running through the TV screen. Digital TV signals, however, are all but invisible unless you have a spectrum analyzer at your disposal. Digital signals appear to the uninitiated IRD as virtually indistinguishable from random noise.

Now, having said the above, it is also fair to point out that digital DTH encryption systems share many of the key aspects that have already been described with regard to analog encryption systems. Pseudo-random binary sequence generators are used to generate the required electronic keys, and the precise synchronization of the encoder and decoder is also an essential requirement. Smart cards and their corresponding conditional access (CA) readers, which are built into all digital DTH IRDs, are also an integral part of the digital encryption equation.

In fact, some digital DTH services use a specialized version of a conditional access system that is already in use elsewhere to encrypt analog satellite TV broadcasts. For example, DIRECTV uses a version of News Datacom's conditional access system that also is employed in Europe for analog TV encryption purposes under the name VideoCrypt. The VideoCipher RS (analog) and DigiCipher II (digital) systems developed by General Instruments also share many common elements.

ADVANCED DEFINITION TV SYSTEMS

In 1983, an organization known as the Advanced Television Systems Committee (ATSC) was created to coordinate the technical details for a replacement standard for the then 41-year-old NTSC standard. In 1987, the FCC appointed an Advisory Committee for Advanced Television Systems (ACATS) to inform the Commission about the technological advances required to produce an advanced television system (ATV) for the United States. ACATS initially evaluated a total of twenty-three different analog-based systems. In 1990, General Instrument submitted the first digital advanced TV system for consideration. ACATS immediately recognized the superiority of digital delivery for any new ATV standard and abandoned the search for an analog alternative that would be backward compatible with the existing NTSC video standard.

In 1990, the FCC responded to the ACATS recommendation to pursue a digital solution by proposing a "simulcast" plan for the introduction of ATV. During the initial phase-in period, existing TV stations would continue to broadcast in NTSC while new ATV stations would broadcast on TV channels that are currently unused within each local TV market.

With the NTSC system, adjacent TV channels are left unoccupied to prevent interference between TV stations operating within the same broadcast area. Digital ATV services, however, can occupy these unused channels without causing interference to analog TV stations. The FCC, therefore, would not need to assign new ATV channel frequencies to introduce ATV services. What's more, there would be total use of the VHF and UHF frequency spectrums assigned for terrestrial TV broadcasting for the very first time.

THE DIGITAL GRAND ALLIANCE

Several organizations introduced digital ATV systems to ACATS for consideration including two digital systems jointly developed by General Instrument and the Massachusetts Institute of Technology, a system jointly developed by Zenith and AT&T, a system jointly developed by Thomson and David Sarnoff Laboratories, and a system developed by Philips. After all systems were fully evaluated, ACATS was unable to designate any one system as inherently superior to the others.

In February of 1993, ACATS encouraged all of the digital system providers to combine the best features from all the systems to create a unified ATV standard for recommendation to the FCC. On May 24 1993, the seven companies announced the formation of a consortium called the "Grand Alliance" to jointly develop a single, "best of the best" ATV system.

Field trials of the new system were held at various locations around the country, including Washington, D.C. At the conclusion of the testing period in early 1996, the final proposal for a new ATV system was submitted to the Commission.

FROM ATV to DTV

On December 24, 1996, the FCC adopted a technical standard for what the Commission now calls digital television, or DTV, that is a modification of the system recommended by the ATSC. The new DTV standard offers a multiplicity of digital TV, audio and data formats, including the broadcast of one or two High Definition Television programs; five or more Standard Definition Television programs at a visual quality better than a NTSC analog signal; numerous CD-quality audio signals; and the delivery of large amounts of data. The DTV standard does not require DTV broadcasters to use specific scanning formats, aspect ratios and lines of resolution. Instead, the DTV standard offers each broadcaster a variety of options from which to choose (see chart below).

DTV promises to enhance the quality of both the sound and the images to be displayed on new DTV-compatible TV sets. DTV images will present up to twice the number of lines (1080 versus 525) over what conventional TV pictures provide today. The video images even have the potential to deliver pictures with a sharpness that approaches the clarity of 35-millimeter film.

Digital Television Standard Display Formats							
Vertical Lines	Horizontal Pixels	Aspect Ratio		Picture Rate Fields per sec.			
480	640	4:3	4:3	60-I	60 P	30 P	24 P
480	704	16:9	4:3	60-I	60 P	30 P	24 P
720	1,280	16:9			60 P	30 P	24 P
1,080	1,920	16:9		60-I		30 P	24 P
I = interlaced scanning P = progressive scanning							

Fig. 7-9. DTV will provide each broadcaster with a variety of formats from which to choose.

The flaws inherent in the old NTSC standard will be eliminated. The DTV standard will accurately portray all the colors of the original image without viewers scrambling to adjust the tint controls on their TV sets. The DTV standard takes advantage of sophisticated digital filtering and forward error correction techniques, so that the DTV set can detect and "mask" out noise, "ghosting," as well as electrical interference from automobiles and electronic appliances. Video "crawl" and other NTSC picture artifacts will also be a thing of the past.

THE DTV SPECIFICATIONS

DTV offers broadcasters the option of delivering their signals in either the standard 4:3 aspect ratio used by today's NTSC TV sets or in a wide-screen, 16:9 aspect ratio that more faithfully reproduces the dimensions of film-based materials. DTV also will use digital "Dolby" audio transmission techniques to broadcast programs in stereo with surround sound.

Pixels and Lines. The DTV standard supports four fundamental arrays of vertical lines and horizontal picture elements or "pixels" that can be displayed on the TV screen: 480 x 640, 480 x 704, 720 x 1280, and 1080 x 1920. Although the NTSC standard is a 525-line system, only 483 of these lines are "active" lines, with the remaining "inactive" lines contained in the vertical blanking internal. Moreover, NTSC generates 756 pixels per line. Therefore, the 480 x 704 and 480 x 640 DTV formats, which are roughly equivalent to NTSC in terms of vertical resolution, are also referred to as "STV" (for standard TV) formats. The 720 x 1280 array has been dubbed the "ADTV" (for advanced definition TV) format because its vertical and horizontal resolution exceed the performance char-

acteristics of NTSC, PAL and SECAM, while the 1080 x 1920 array is an HDTV (high definition television) format.

Field and Frame Rates. Field rates of 60, 30 and 24 fields per second are available. The 60 and 30 fields per second best accommodate video source material using interlace scanning, while the rate of 24 frames per second is advantageous for the transmission of all film-based source materials using progressive scanning. Media resources that use progressive scanning, such as film-based materials, differ from video resources using interlace scanning in that each line of an image is presented in sequence.

Video Compression System. DTV uses the MPEG-2 specification as the basis of its own compression system. DTV takes advantage of the layered structure of MPEG-2. One layer can transport an STV signal to less-expensive DTV sets, while at the same time additional layers can transport signal enhancements that will allow more expensive ADTV or HDTV sets to display higher-resolution images from the same digital TV broadcast. MPEG-2 data packets also provide for the transmission of virtually any combination of video, audio and data information. One major difference between an MPEG-2 DVB-compliant signal and a DTV signal is that the former uses a modified version of MUSICAM for the creation of CD-quality digital audio while DTV will use the 5.1 channel Dolby AC-3 surround sound system.

THE ROAD AHEAD

On April 3, 1997, the FCC formally established a schedule for a phased transition from NTSC to DTV beginning in 1999. All TV broadcasters must provide a free digital video programming service that is at least comparable in resolution to their existing NTSC service and also aired during the same time period. The FCC's transition timetable requires that the affiliates of the major four networks in the country's top ten TV markets be on the air with a digital signal by May 1, 1999. Affiliates of the major four networks in markets eleven through thirty must be on the air by November 1, 1999. All other commercial and non-commercial stations must complete their transition to DTV by May 1, 2002, and May 1, 2003, respectively. (The top ten markets include 30 percent of all American TV households, while the top 30 markets include 53 percent of all U.S. TV households.)

Released on April 21, 1997, the FCC's "DTV Table of Allotments" calls for the eventual location of all DTV channels in a core spectrum of VHF (2 to 13) and UHF (14 to 51) channels that are technically most suited to DTV broadcasting. At the end of the transition period, each broadcaster must return the frequency spectrum that it currently uses to transmit analog TV signals. UHF TV channels 60 through 69 (764 to 806 MHz) will be returned prior to the end of the transition period so that the spectrum can be assigned for other purposes, such as fixed and mobile services for public safety use.

The Commission has set a target date of 2006 for the total phase out of the NTSC system. The FCC, however, will periodically review this date in light of its evaluation of the transition to DTV.

In the near future, DTV manufacturers will begin to market digital set-top boxes that will permit a terrestrially transmitted DTV signal to be displayed by any NTSC TV set. A variety of DTV sets also will appear soon in the marketplace, offering the satellite service provider with an opportunity to sell and install DTV sets along with satellite receiving hardware. The new DTV sets featuring a 16:9 aspect ratio also will compel the digital DTH service providers to begin offering ADTV and HDTV broadcasts. The advanced capabilities of DTV will give terrestrial TV broadcasters and cable TV systems the opportunity to more effectively compete with digital DTH service providers. The availability of the new DTV sets in the marketplace also will allow DBS operators such as DIRECTV to begin broadcastng movies in a high-definition, wide-screen format as early as 1999.

The old Chinese curse "May you live in interesting times" in this case actually should be regarded as a blessing in disguise. The transition to DTV will provide all of us with new opportunities and greater freedom of choice as we enter the first few years of the new millennium.

SATELLITE INSTALLATIONS

Although the vast majority of satellite TV purchasers leave the installation of their home satellite TV systems to experienced technicians, an increasing number of individuals are electing to install their own systems, especially when the antenna size for a digital DTH system is managable for nearly everyone.

For the larger C-band dish systems, a popular option is to undertake the bulk of the installation yourself, including mechanical parts such as assembling the dish and running the cables. A technician can then be hired to tackle the more complicated installation steps, such as adjusting the dish so that it will properly track the satellite arc or programming the satellite IRD. The satellite technician can complete both of these steps quickly and easily once the equipment is in place. If you have not previously operated a satellite TV system and aren't sure what to expect, the technician also can walk you through the basic operating procedures, explain all IRD or receiver functions, and give you confidence that you will be getting the most you can out of your new system.

Your satellite retailer should have a technician available or be able to recommend one to you. He or she also should supply you with the information that you will need in order to locate the satellites that are available from your location: azimuth, elevation, and declination angles. If not, you can derive this information from the charts provided later in this chapter, or visit my web site at http://www.mlesat.com for a free download of a computer program that will do the calculations for you.

Your antenna and electronic components will come with assembly and operating instructions. The following step-by-step guide will walk you through the installation and explain the various procedures that you will perform along the way. Whether you decide to tackle the installation on your own or hire a technician, an understanding of site selection criteria, cable connections between components, and dish aiming techniques will improve your operating skills and increase your enjoyment of the system.

SITE SURVEYS

Satellite signals are microwaves that travel in a straight path along the line of sight. Since all geostationary satellites are positioned over the Earth's equator, if you want to receive them from the Northern Hemisphere your antenna must have an unobstructed view of the southern (or northern for sites located below the equator) sky. So, before you go running out and spending hundreds or even thousands of dollars for a DTH receiving system, you should be sure there are no tall buildings, trees, power poles, or other substantial obstacles to prevent the signals from reaching your dish.

Your local satellite retailer usually can conduct a site survey at your location. Often the fee for this service is deducted from the eventual purchase price of a system. If the site survey takes place in late autumn or winter, you may need to take into account the foliage of nearby trees and bushes that will be in place during the spring and summer months. Future construction plans of neighbors or nearby businesses also must be considered; any new buildings could be in a position to affect your reception.

You can conduct your own preliminary survey by standing in your backyard and looking toward the south (or north for sites located below the equator). If you have a clear and unobstructed view of the sky from the southeast to the southwest (or northeast to northwest for stations south of the equator), then you are in luck. However, the view from many installations, especially those in urban or suburban areas, will be blocked in some direction. At these restricted sites, some satellites may not be in view. To determine which satellites will be visible from your location, you will need to learn some basic techniques for locating the satellites.

AZIMUTH AND ELEVATION

Azimuth and elevation are the two basic coordinates used to determine each satellite's position in the sky. The azimuth coordinate represents the bearing of the satellite from your location, while the elevation is the angle at which the dish looks up at the satellite. Every satellite within view of a specific installation has its own unique pair of azimuth and elevation coordinates. Given the site's latitude and longitude, the coordinates for any satellite are fairly simple to calculate.

Once you know the azimuth coordinates for a specific satellite, you can use a compass to point the antenna in the right direction. To achieve the required accuracy, you must correct the compass readings for true north. You can call up the control tower at the nearest airport to find out the correction factor for your location. If true north is east of magnetic north, the correction factor must be subtracted from the compass readings. When true north is west of magnetic north, the correction factor is added. All compass readings should be taken out in the open, well away from large metal objects or overhead AC power lines and transformers.

From any location, the satellite nearest to true south (180 degrees from true north) will be at the highest elevation point of all the satellites for sites north of the Earth's equator. The further away the satellite's bearing is from true south, the lower its elevation will be. At some point, the elevation of satellites to the extreme west or east of your location will fall below the horizon, preventing reception. That is why satellite dishes in the Americas cannot view most European satellites, and vice versa.

An inclinometer is used to measure the degrees of elevation, or "tilt," of the dish. Inclinometers, some-

Fig.8-1 & 8-2. Azimuth and elevation coordinates for locations north (top) and south (bottom) of the Earth's equator.

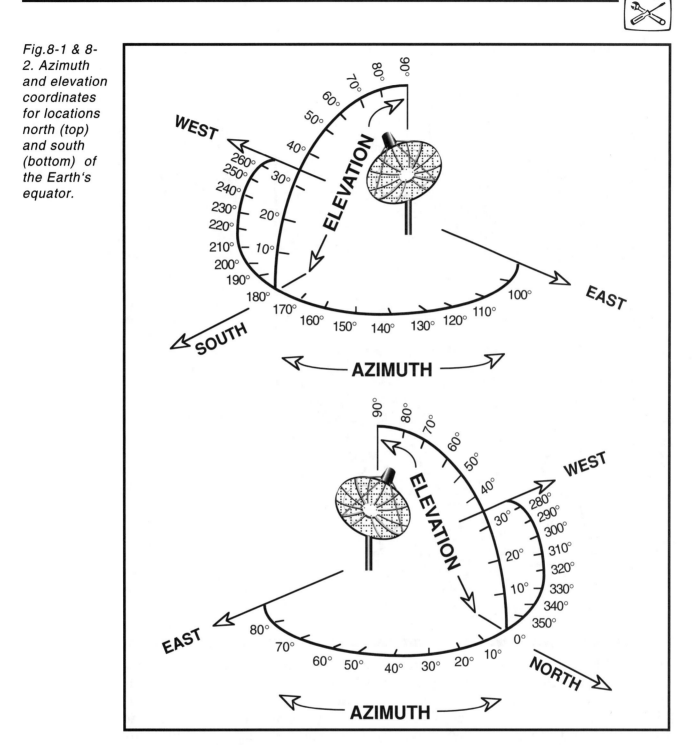

THE AZIMUTH VALUES MUST BE ADJUSTED TO CORRESPOND WITH LOCAL TRUE NORTH/SOUTH BY ADDING (OR SUBTRACTING) A MAGNETIC CORRECTION FACTOR WHICH CAN BE OBTAINED BY CALLING THE LOCAL AIRPORT CONTROL TOWER. THE ELEVATION ANGLE FOR ANY SATELLITE IS ALSO REFERRED TO AS THE INCLINATION ANGLE, WHERE THE HORIZON IS REFERENCED AS A ZERO DEGREE ANGLE AND THE PERPENDICULAR ANGLE, OR STRAIGHT UP, IS NINETY DEGREES. EACH SATELLITE HAS ITS OWN UNIQUE SET OF AZIMUTH AND ELEVATION COORDINATES. THE NUMERICAL VALUES FOR EACH SET OF COORDINATES WILL VARY, DEPENDING ON THE PRECISE LATITUDE AND LONGITUDE OF THE RECEIVING SYSTEM'S LOCATION ON EARTH.

UNIVERSAL AZIMUTH-INCLINATION LOOK ANGLES

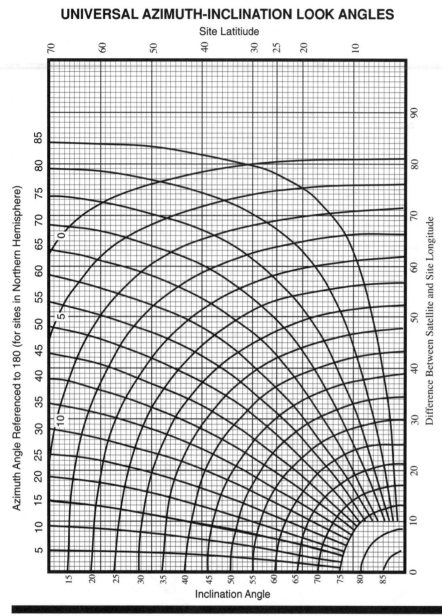

1. Subtract site longitude from desired satellite longitude and plot this value on the bottom scale of the chart.

2. Plot the site latitude using the scale on the right side of the chart.

3. At the point where A and B intersect, determine the elevation angle and difference in azimuth using their respective graphs.

4. If the value obtained in step 1 is a positive number, add the difference in azimuth from step 3 to 180 to obtain the azimuth bearing from true south. If the value is a negative number, subtract the difference in step 3 from 180 degrees.

Fig. 8-3. Universal Azimuth/ Elevation angle chart.

times called "angle finders," are readily available from satellite TV industry sources. Hardware shops and department stores may also sell inclinometers to carpenters who use them to measure roof pitch angles.

An inclinometer also can be used during the site survey to determine whether obstructions would block out any of the available satellites. Position the inclinometer on a level surface at the center of the site, and line it up with the bearing, or "azimuth," of the satellite under consideration. Then you can tilt the inclinometer back until it indicates the desired satellite's elevation. By sighting along the edge of the inclinometer, you should be able to tell if any obstacles would be between the site and the satellite.

If you determine that there will be obstructions that will interfere with your reception of a single satellite or

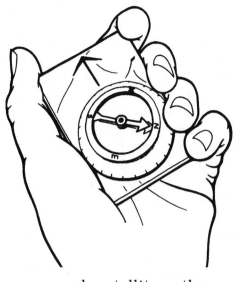

Fig. 8-4. An inexpensive compass.

Fig. 8-5. Do not be confused by two edged inclinometers. You will get one reading when you hold it against one edge and another reading when using the other edge. Use the edge underneath the "0" degree mark on the inclinometer's scale.

even several satellites, then you should remove these obstacles. In some cases the elimination of a troublesome tree branch may suffice. When the obstruction is part of a nearby building or hill, the proposed dish site sometimes may be moved from ground level to a more advantageous rooftop location. Roof and wall mounts are readily available for small digital DTH antenna installations. For large aperture C-band or Ku-band installations, a tall pole that extends from the

ground up the side, and anchored to, the house can allow the antenna to peer over the roof and receive a clear view of the satellites without actually being anchored to the roof itself.

C-band satellite receiving systems can receive interference from land-based telephone microwave stations or airport, shipyard, and military radar installations. This can restrict or even preclude satellite reception. If your home is between two microwave links or near a radar installation, reception may be affected. Ku-band digital DTH systems seldom encounter terrestrial interference (TI) problems.

If you are contemplating the installation of a C-band receiving system, the best way to check for interference is to have your local satellite TV retailer perform an on-site demonstration with a portable dish. To detect all potential interference, the test should be performed right at the spot where the proposed dish will stand. A difference of only a few feet can be critical.

Any on-site demonstration should be made on a weekday, when telephone and other microwave traffic is at its peak. If you are in a business area, test during business hours; in a residential area, check it later in the day. In the event that some interference is present at your location, there are filters available that can often eliminate it. Information is provided on microwave filters in the chapter on troubleshooting.

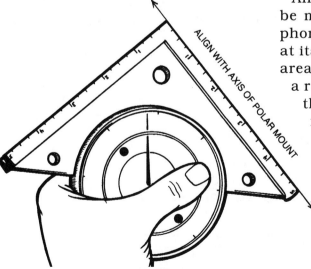

ALIGN WITH AXIS OF POLAR MOUNT

SMALL DISH INSTALLATIONS

Several of the high-power DBS service providers offer complete home installation kits for their small dish systems. With the proper tools and a modicum of enthusiasm and effort, it should not be difficult for the hobbyist to install a small, fixed satellite antenna for receiving TV programs from an individual satellite or group of satellites located at one orbital location in the sky. In fact, some of the available self-installation kits include a videotape that will take you through all of the required steps. In many cases, the most important step will be to select and install the appropriate antenna mount for your particular location and living situation.

Selecting the Antenna Mount. Mounts are available in a variety of styles to accommodate installation of the antenna on the ground, attached to an outside wall, or under the eaves of the house. Roof mounts are available that either penetrate the surface of the roof and attach to the building's rafters or mount directly on the top of a flat roof. Above all, select the type of mount that can be installed where the antenna will have a clear and unobstructed view of the satellite of choice throughout the year.

Installing the Antenna Mount. Once you have securely fastened the antenna's mount to the building or a

The ESSENTIAL SATELLITE INSTALLATION KIT
(What You'll Need Before You Start)

Tools
Compass & inclinometer
Post hole digger & pointed shovel (ground mounted systems only)
Screwdriver, tape measure, crimp tool
Wire strippers, wire cutters, razor knife
Electrical power cord with multiple a.c. outlet
Large and small adjustable wrenches and at least one 7/8 inch wrench
Drill with large bit (masonry for concrete block or wood bit for frame homes)

Supplies
Plastic wire ties
Voltage surge protector
Quick setting concrete mix
Connector sealer (Coax-seal, etc.)
Current Satellite TV Program Guide
Ground rod, clamp, and copper ground wire
"F" connectors (RG-56 type for direct burial cable)
Direct burial cable with two coaxial cables, one servo, and one actuator motor cable
Steel pole of the correct length and outside diameter for your dish's mount
Plastic plate to cover the hole in the wall where the cable enters the house

---

Fig. 8-6. The essential satellite installation tool kit.

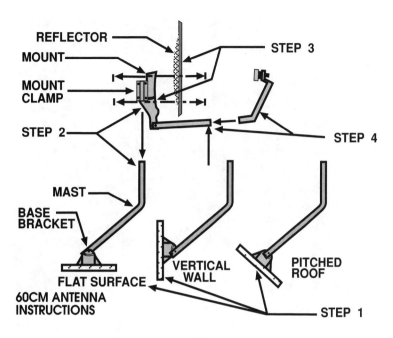

Fig. 8-7. Instructions for various ways that you can install a 60cm offset antenna.

REFLECTOR
MOUNT
MOUNT CLAMP
STEP 2
MAST
BASE BRACKET
FLAT SURFACE
60CM ANTENNA INSTRUCTIONS
VERTICAL WALL
STEP 3
STEP 4
PITCHED ROOF
STEP 1

supporting pole, mount the antenna as indicated in the assembly directions. Do not fully tighten down those bolts which anchor the antenna at a specific position of elevation and azimuth until you have adjusted the antenna pointing to maximize signal reception. Mount the feedhorn and LNF onto the feed support supplied with the antenna.

Running the Cables. A coaxial cable will have to be installed between the receiver's location in the home and the low noise feed (LNF) out at the antenna. For most installations, this coaxial cable will carry both the incoming signal and a switching voltage that will command the LNF to switch between opposite senses of polarization. Some systems may require a two- or three-wire cable to connect the indoor receiver to the polarization connections for the LNF.

Cable clips can be used to fasten the system cables to the side of the building. Run the cable a bit below and then back up to the hole into the building, thus providing a "drip loop"

to prevent rain from running down the cable and into the building. Quality crimp-on "F" connectors should be used at each end of the coaxial cable, although the reusable twist-on F connectors are gaining in popularity.

The power for the LNF also is supplied by this coaxial line. Whenever you plug the receiver into the home's AC wall receptacle, electricity is being sent up this line. To avoid a short circuit that would blow the receiver's fuse or circuit breaker, or even damage the LNF, always unplug the receiver whenever connecting or disconnecting the system's coaxial cable.

Peaking Antenna Performance. Before connecting the cable to the LNF, determine which method you will use to aim the antenna at the satellite of choice. One antenna peaking method used when installing analog satellite TV receiving systems is to view the available TV services from this satellite while adjusting the antenna's alignment. You can then fine tune antenna pointing until all stray impulse noise or "sparklies" have been eliminated for all available channels. You can get close to maximum system performance using this method.

The above-mentioned method is not at all appropriate when receiving digital DTH services. Receiving digital signals is an all or nothing proposition: you will either receive a perfect picture or no picture at all. There is no intermediate state of marginal reception that you can use as a starting point. In this case, it is essential that you use some kind of signal tuning meter to detect the presence of signal and then peak the antenna for maximum performance.

The use of a tuning meter also is a good idea when receiving standard analog TV signals. Rain, fog, snow or even rain-filled clouds passing overhead can reduce the intensity of any Ku-band satellite signal reaching your antenna. If you tune your antenna by watching the TV screen, you may receive a wonderful picture at the time of installation, when there may be enough signal to exceed the receiver's threshold rating, but subsequently lose the signal whenever bad weather occurs.

Keep in mind that the meters installed on most satellite receivers and IRDs are not nearly as sensitive as external peaking meters. That is why it is so important to peak system performance at the outset, so that the amount of signal reaching the receiver exceeds the unit's threshold level. This signal margin will then be there to compensate for any signal degradation that occurs during adverse weather conditions.

Most set-top boxes provide an on-screen graphic which displays relative signal strength on your TV screen. They also feature a built-in tone generator that generates a shrill whistle that rises and falls in pitch to indicate changes in signal strength. This is a very useful feature to have because you don't want

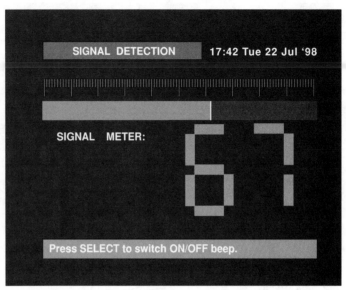

Fig. 8-8. The IRD that comes with all digital DTH receiving systems has a built-in signal meter that you can display on your TV screen.

to haul the IRD and TV set out to the dish in order to determine what's going on, especially when peaking an antenna that is mounted on the ground or a flat roof. Turn up the volume on the TV set so that you can hear the tone while working outside at the antenna. As you adjust the antenna slightly to the right/left and up/down, the audio tone will assist you in finding the antenna's "sweet spot."

If you are forced to move the IRD and TV outside, an extra length of coaxial cable (and possibly the feedhorn cable) will be needed to temporarily connect the IRD to the LNF. It is impractical to lug the IRD and/or TV up a ladder to peak an antenna mounted on the side of a building or under the eaves. What's more, a dedicated signal-strength meter also will be more accurate and sensitive than any other meter supplied inside the IRD. The acquisition of a high-quality, in-line, signal-strength meter is a must for the professional installer.

One word of caution concerning the use of any signal strength meter: you must always be sure that you are receiving the correct satellite. There are many satellites in the skies overhead these days, and you don't want to do a lot of work only to find that you are anchored onto the wrong one! Scan through the available channels at the beginning and again at the end of the installation process to ensure that you are getting the desired satellite TV services.

The signal strength meter may even supply the DC voltage needed to power up the LNF and therefore can connect directly to the LNF. Signal meters that do not supply power to the LNF will have to be inserted in the coaxial line running from the receiver to the LNF using a special "tap" connector and two coaxial jumper cables to connect the signal meter to the tap and LNF. The tap supplies three connectors: one connects to the LNF, the second connects to the receiver or IRD, and the third supplies the connection that allows the meter to "tap" into the line and make the required signal strength measurements. Once you have completed the antenna alignment, you can tighten down all of the mounting bolts to prevent strong winds or rain from re-pointing the antenna for you.

Setting the Polarization. The LNF support that comes with the antenna holds the LNF at a fixed position towards the front of the antenna. If this support has an adjustment for setting the distance between the LNF opening and the surface of the dish, try moving the feed in and out in tiny intervals of a fraction of a centimeter to see if the signal strength increases.

All high-power DBS satellites use opposite senses of circular (right-hand and left-hand) polarization to transmit a greater number of channels from a single orbital position. Medium-power Ku-band DTH satellites will use two senses of linear (horizontal or vertical) polarization to accomplish the same thing. To receive both senses of polarization, the LNF electronically switches its sense of polarization by 90 degrees. For most effective reception of the digital DTH services coming from a satellite using linear polarization, you will need to peak one of the two senses, either vertical or horizontal. The LNF will then be able to automatically switch to the other sense to receive those signals at maximum signal strength.

If you are receiving signals from a high-power DBS satellite, you will not have to be concerned about this. An LNF that has been designed to receive signals using circular polarization will receive either the right-hand or left-hand senses at maximum signal strength at all times. With circular polarization there are no intermediate states between the two orthogonal (at right angles) senses.

LARGE DISH INSTALLATIONS

If you proceed with the installation of a large aperture antenna on your own, you will need to construct a foundation for the mount. In the case of a pole-mount antenna, a steel pipe is placed in a hole in the ground and embedded in concrete. Because it will take a few hours for the concrete to harden, set the pole the day before you construct the antenna. You also can buy quick drying concrete products that will set up in less than an hour. Be sure to use a level to ensure that the pipe is "plumb" before the concrete sets.

Other types of dish mounts will require a concrete pad for the foundation. Concrete piers extending well below any frost line should be incorporated into the pad design. Certain types of antenna mounts merely need a level surface. The dish manufacturer supplies all of the detailed instructions that you'll need.

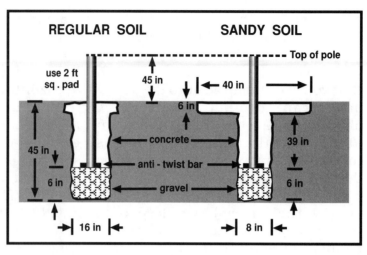

Fig. 8-9. Ground-based pole mount installation instructions.

For rooftop installations, the roof structure must be able to support the weight of the dish as well as withstand the uplift forces of several thousand kilograms resulting from moderate and high-speed winds. After all, nobody wants the roof torn off and the dish flung through the sky.

Roof top installations are more susceptible to microwave interference than installations located at ground level. A site survey with a spectrum analyzer, or an on-site rooftop demonstration with a small portable dish, is an absolute necessity in any area where microwave interference is a major potential problem. If several of your neighbors have antennas, you can check with them to see if they are experiencing any interference problems.

ASSEMBLING THE DISH

The dish antenna comes with a detailed set of instructions covering all the essential construction information. When you first receive the dish, check to see that all parts listed in the manual are present and accounted for. Read over the instructions to see what tools and other materials will be required. If assistance during assembly is recommended, arrange to have a friend or two available to help lift the dish onto the pole.

On the evening prior to assembly, check the local weather forecast to ensure that rain or heavy winds will not interrupt your efforts. Begin assembly early in the day to allow enough time to complete the project.

Take care during the assembly to

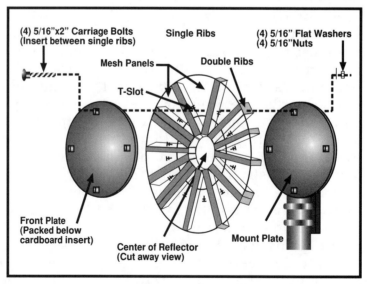

Fig. 8-10. Mesh antenna rib assembly instructions.

Fig. 8-11. Assembly of a quad mesh reflector.

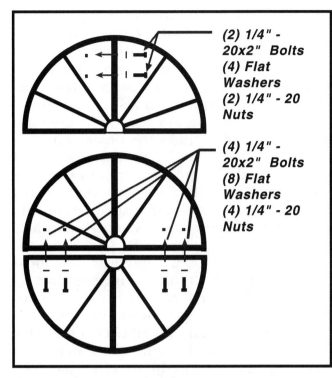

(2) 1/4" - 20x2" Bolts
(4) Flat Washers
(2) 1/4" - 20 Nuts

(4) 1/4" - 20x2" Bolts
(8) Flat Washers
(4) 1/4" - 20 Nuts

ensure that you get the most accurate curve possible. Many popular dishes today are comprised of four sectional pre-assembled panels called "quads" which bolt easily together. You'll want to have the dish fully assembled and anchored before leaving it unattended.

MOUNT ALIGNMENT

The pivotal axis of a polar mount antenna must be accurately aligned to true north. Even when this alignment is done with an engineer's transit, slight inaccuracies in the measurement may result, causing uneven tracking of the geostationary satellites. Corrected compass readings may be even further off the mark. While compass readings may be useful during the site survey for locating a clear view of the satellites, the real test of your polar mount's alignment is the video reception that you obtain.

When mounting the antenna onto the pole, tighten the bolts just enough to hold the antenna in place. Tightening the bolts down firmly on the pole should be one of the very last things you do, and should be done only after you have gotten the dish to track the satellites properly. Tightening beforehand can dimple the pipe, making it more difficult to make subtle adjustments in alignment.

MOUNTING THE FEED & LNB

Extreme care should be taken when bolting the LNB to the feedhorn. Do not touch the probe inside the mouth of the LNB. Any grease or dirt covering this probe will adversely affect the LNB performance. Also make sure that the neoprene gasket that goes between the feedhorn flange and the mouth of the LNB is "seated" properly in the groove provided. Otherwise, moisture can seep through this opening and disrupt your reception. If you are installing an LNF as a part of the system, you will not have to worry about any of these items.

The center of the feed must be supported exactly at the focal point of the dish. The focal distance between the center of the dish and the mouth of the feedhorn will vary between different models of antennas, depending on whether the manufacturer has used a deep or shallow dish design. The exact distance will be provided in the instructions.

There are two feed support styles available, one of which will be supplied with your antenna. A "button hook" support is a single piece of tubing which extends from the center of the dish outward. The buttonhook support, which is curved so that it resembles a hook, allows the feed to be mounted looking back at the center of the dish. Guy wire kits are available for this type of support that will provide the additional stability required for reception of both C- and Ku-band signals. Ku-band signals are much higher in frequency and therefore are beamed Earthward in much smaller wavelengths. Consequently, the antenna curvature and location of the feed for Ku-band reception must be much more precise than what C-band systems commonly require.

It is recommended that you check to ensure that the feed is centered over the dish by measuring from the lip of the antenna to the edge of the feedhorn opening at four equally spaced intervals around the antenna's rim. This is important when you are using guy wires to anchor the button hook support. If the feed is centered, these measurements will all be equal. If they are not equal, you will have to adjust the tension of the guy wires until the feed is properly centered.

The second type of feed support uses a multi-legged structure to hold the feedhorn and LNB (or LNBs for dual-band systems). These are made up of three ("tripod") or four ("quad") straight, equal length pieces of aluminum or steel. Quad supports are inherently more stable than buttonhook supports, offering better Ku-band reception. When mounted at the recommended locations on the antenna's surface, these supports should accurately position the feed at the correct focal length— the distance between the center of the dish and the opening of the feedhorn. The correct focal length for your dish is provided in the manufacturer's assembly manual.

Once your system is up and running, you can fine tune the focal length for its optimum position by moving the feed in and out in small increments while watching the IRD's signal strength meter. This is easier to do with a buttonhook than a quad support. This adjustment is particularly important if you are having trouble receiving Ku-band satellite signals.

You can compute the focal length if you know the diameter of the dish along with its f/D ratio. Focal length = antenna diameter times the f/D

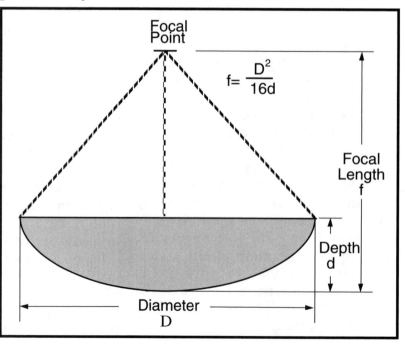

$$f = \frac{D^2}{16d}$$

Focal Point

Focal Length f

Depth d

Diameter D

Fig. 8-12. Determining the focal length of any dish.

ratio. For example, the focal length of a 10 foot antenna with a f/D ratio of .45 equals 10 x .45 = 4.5 feet (54 inches).

To determine the antenna's diameter, measure across the surface of the dish from one side to the other. The radius equals one half the diameter. The depth of the dish is the distance from the center of the dish to the plane of the rim. Stretch a string across the antenna's rim so that it crosses in the center of the dish. The depth will be the distance from the antenna's center to the string.

Many feeds today have adjustable scalar rings. These feeds can be broken down into two parts: a round flat "scalar" plate with concentric circles on its surface and the waveguide onto which the LNB is mounted. This waveguide (technically referred to as an "orthomode coupler") fits into the center of the scalar plate and can be adjusted inward and outward.

The distance that the waveguide extends beyond the surface of the scalar plate must be set to correspond to the f/D ratio of the antenna. Consult the manufacturer's assembly directions or use the formula provided above to determine the correct f/D ratio of your antenna.

The waveguide may be marked to indicate the various f/D ratio settings. Alternatively, the feedhorn may come with an adjustment gauge for setting the correct location of the scalar rings.

Another thing to check: the plane of the feed opening should be the same as the plane of the rim of the dish. You can use your inclinometer to check to be sure that both the feed opening and the antenna's rim are parallel with each other.

RUNNING THE CABLE
In most cases it will take three sets of wires to hook up your system. The coaxial cable runs from the LNB to the receiver. A three-conductor wire connects the servomotor of the feedhorn to the servomotor control on the back of the receiver. A four or five conductor wire connects the actuator motor at the antenna to the control terminals on the back of the receiver or IRD. We will examine each of these wires in detail in the following sections.

The best way to acquire these wires is to purchase an all-in-one direct burial satellite cable from your retailer at the time you purchase your system. Bury the cable in a trench deeper than the local frost line and you are ready to go.

Purchase more than enough cable to complete the run. The cable may run all the way from the antenna feedhorn, down the pole, in the ground from the pole to the house, up the wall, through the attic, and down an interior wall (which is not always, but oftentimes, the case). Leave enough cable to easily reach the back of the indoor unit. If you do come up short, buy an extension cable with pre-attached male F connectors and a double female barrel F connector to lengthen the line.

Before connecting all of these cables, however, self-installers should consider temporarily using a short piece to connect the receiver and TV to the outdoor electronics right out at the dish. This method for aligning the dish allows you to make tracking adjustments to the antenna while viewing TV signals

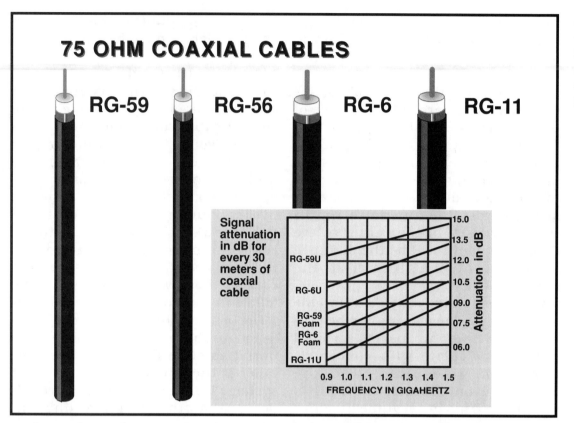

75 OHM COAXIAL CABLES

RG-59   RG-56   RG-6   RG-11

Signal attenuation in dB for every 30 meters of coaxial cable

Fig. 8-13. Signal attenuation for 75 ohm coaxial cables.

and watching the receiver's built-in signal level indicator. If you are going to use a technician during the final stages of the installation he or she will have a portable meter to use at the dish. Two people also can talk back and forth via walkie-talkies or portable phones while one makes adjustments to the dish and the other keeps an eye on the quality of the reception.

Direct burial satellite cable is available containing either one or two coaxial lines. The direct burial cable with two coaxial lines is essential if you intend to use either a hybrid feed (which requires one C-band and one Ku-band LNB) or a dual C-band feed (which requires two C-band LNBs). We recommend that you purchase the direct burial cable with two coaxial lines even if you initially intend to use only a single LNB. The spare coaxial line will then be in

place in the event that you ever want to upgrade your system or encounter problems with the primary coaxial line.

If you intend to have more than two LNBs at the dish, you will need to run separate coaxial lines for each of them. For example, multiple receiver systems with feedhorns which support two Ku-band LNBs and one C-band LNB, or two C-band and two Ku-band LNBs, will need multiple coaxial lines. We go into detail about multiple receiver systems in a later section of this chapter.

COAXIAL CABLES & CONNECTORS

You already may be familiar with a shielded wire cable called coax ("koh-axe," from coaxial) most often used for connecting your TV set to the home's master antenna system. Coax is made up of an inner wire covered with a plastic or foam sheath, and

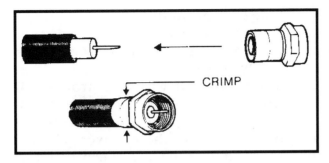

Fig. 8-14. F connectors.

CRIMP

an outer mesh that is in turn surrounded by an outer plastic covering.

A single coaxial cable is used to carry the signal from the outdoor electronics to the indoor unit. A shorter piece of coax also is used to connect the receiver to the antenna input on the TV. If the programming is to be viewed on other TV sets, this cable will first go to a "splitter" and then on to the various TVs.

A unique type of cable TV connector, called an "F" connector, is crimped onto each end of the coaxial cable. It mates with complementary connectors on the LNB and set-top box. Your local electronic supply house can provide you with the special tool used to crimp "F" connectors onto the cable. Or you can purchase standard lengths of cable with the connectors already installed. The quality of the F connector is important as some cheaper F connectors break when crimped, providing an additional point for moisture to gain entry as well as a less reliable connection.

When screwing an F connector onto mating connectors on the back of the IRD or LNB, you should take care to avoid bending or breaking the cable's inner conductor, thereby shorting out the connection. Right angle F connectors also are available that can be used whenever space limitations prevent a straight-on connection.

The block IF signal coming from the LNB is no longer a microwave signal. That's why a relatively inexpensive and readily available coaxial cable of small diameter can be used to carry the signal to the indoor unit. Satellite TV cable has a characteristic impedance rating of 75 ohms. The type of cable used by CB radios and other two-way radio equipment is 50-ohm coax and therefore is not a suitable replacement for the 75-ohm coax used in all satellite TV installations. Be sure that the cable you buy is 75-ohm coax.

There are several different kinds of 75-ohm coaxial cable available. RG-59U coax can be used to span distances of up to 100 feet. For longer lengths, lower-loss RG-6 or RG-11 cables are used. Direct burial satellite cable contains one or two spans of RG-6. Since RG-6 is slightly larger in diameter than RG-59, it also requires a slightly larger F connector. To span distances of several hundred feet, special UHF line amplifiers with +10 or +20 dB gain also may be necessary to compensate for the amount of signal attenuation along the length of cable.

As the block IF frequency range produced by the LNB extends to 1450 MHz or higher, the losses in most types of 75-ohm coaxial cable can be substantial. Use a high quality coax from a major manufacturer from Europe or the United States.

The power required to operate the LNB (and any line amps if used) is supplied by the indoor unit and sent to the LNB via the center conductor of the coaxial cable. The power stays on even if the IRD is turned off. This keeps the LNB at a more consistent temperature and prevents moisture from condensing inside it. Also available are watertight devices that snap onto the outside of the LNB's F connector to keep moisture out.

The indoor unit should be unplugged from the AC wall receptacle before connecting or disconnecting the coax cable from the LNB or the IRD. This eliminates the chance of a short circuit across the coaxial connections.

The connection to the LNB (or LNBs for dual-band systems) also should be weatherproofed to keep moisture out. Flood the connector's interior with a dielectric silicone sealer or wrap the connection with a sticky waterproof compound such as Coax-Seal. It is also a good idea to use a plastic LNB/feedhorn cover to give your outdoor electronic components added protection from the outside environment.

WIRING THE FEEDHORN

The feedhorn line of the direct burial cable is comprised of three color-coded 22 gauge (or larger) stranded wires. These wires also are shielded and jacketed. The three wires provide power, pulse, and ground connections for the feedhorn. Each of these wires connects to corresponding terminals on the back of the receiver. The wires are color-coded to help identify them when connecting to the three servomotor wires at the feed (usually red for power, white for pulse, and black for ground).

A stranded wire is used because it is more flexible and will not break as easily as a solid wire. The shield, an aluminum foil wrapped around all three wires, keeps impulse noise from entering the line and giving false pulses to the receiver. The receiver uses pulses to keep track of the position of the feedhorn pick-up probe. You therefore can adjust, or "skew," the position of the probe and program the optimum polarization for any given satellite transponder into memory.

The feedhorn servomotor rotates the pick-up probe, which swings back and forth while switching between the horizontal and vertical polarization transponders (odd and even channels). Keep in mind that there are limits to the pick-up probe's clockwise and counterclockwise movements. The feedhorn must be aligned on the antenna so that the probe can swing the 90 degrees from horizontal to vertical (or left-hand to right-hand circular) polarization without reaching the limits of its

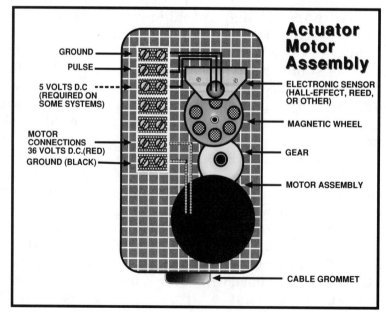

Fig. 8-15. Antenna actuator wiring diagram.

travel. Several manufacturers include a directional guide with their feedhorns to show the proper alignment of the feed when installed on the dish. If you find that you cannot skew the probe beyond a good picture on both the odd and even channels on all satellites, you will need to loosen the clamp that holds the feed onto its support and physically rotate the feedhorn until it is possible to do so.

WIRING THE ACTUATOR

The direct burial cable's actuator line is comprised of five stranded wires. Two 14- or 16-gauge stranded wires are used to power the motor and three color-coded 22-gauge shielded wires connect to the sensor. These actuator wires should be connected to the appropriate terminals on the back of the receiver (or a separate actuator power supply).

Like the servomotor wires, the three shielded motor sensor wires also provide power, pulse, and ground. The vast majority of actuator motors do not require power to be hooked to the sensor. Look inside the actuator housing. If there are only two wires connected to the sensor, then hook up pulse and ground to their respective terminals. If there are three wires connected to the sensor, then hook the 5 volts from the receiver to the red wire on the actuator's sensor, and pulse and ground interchangeably to the other two sensor wires.

The two large stranded wires connect to the large wire terminals at the actuator motor and to the motor's "1" and "2" wire terminals on the back of the receiver or power supply. Now try to move the dish to the east or west. If the dish moves in the direction opposite to the one intended, reverse the wires connected to the motor's main wire terminals.

Some satellite IRDs have an external power supply that puts the large transformer outside of the receiver chassis. This reduces the receiver's size as well as its operating temperature. The power supply is actually a large transformer that turns the local main's AC supply voltage into 24 to 36 volts DC to power the actuator's DC motor.

This outboard power supply can be placed behind the TV set or at some other out-of-the-way place. Moreover, in a two-IRD system you will not need to purchase a second power supply because only one receiver can control the antenna's position. The receiver with the power supply is then called the "master," while any additional receivers are called "slave" units.

Water intrusion can be a big problem with actuator motors. Be sure to run the wires through the rubber grommet supplied with the actuator to seal off the opening. Also mount the motor so the drain holes are on the bottom to allow any moisture to run out.

GROUNDING THE SYSTEM

If your home's AC electrical ground is close to the dish, use a No. 10 AWG or larger solid copper ground wire to connect it to the pipe supporting the antenna. If your dish is physically removed from your home, pound in a separate grounding rod and use a No. 10 AWG or larger solid copper ground wire to connect the pole to it. You should also install an antenna discharge unit or ground block, a passive electrical device that connects in-line between your outdoor electronics and the indoor satellite receiver. To work properly, the ground block should be connected to a ground rod or to the AC ground of the house. For added protection prior to a lightning storm, or whenever your system is left unattended and unused for long periods of time, you should first unplug the satellite receiver from the AC wall outlet and then disconnect the incoming coaxial cable(s) from the IF input of the receiver. After the storm has passed, reconnect the coaxial cable to the receiver's IF input port before plugging the receiver back into the AC wall outlet. This will help prevent damage due to lightning or related power surges.

There are quick disconnecting adapters available that allow you to quickly and easily disconnect all of your indoor components from the rest of the system. You also should use a surge protector on your AC line to prevent voltage surges or spikes from setting your receiver aglow.

ATTACHING THE ACTUATOR

Your dish will have either a horizon-to-horizon mount or a mount that requires an actuator arm. For "horizon" mounts, the motorized housing will attach directly to a mating flange on the mount. Since there are several steps involved in installing motorized actuator arms, I will describe these steps in detail. The method used to set the motor's programmable limits is the same for

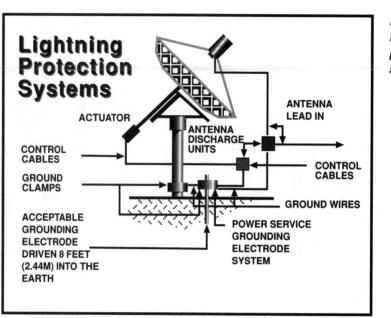

Fig. 8-16. Lightning protection systems.

either style mount.

Before you begin to install your actuator arm, prop the dish up a few degrees with a block of wood or something else that won't damage the antenna. Attach the arm to the dish and the mount as indicated in the manufacturer's assembly directions. The arm normally attaches on the west side of the dish on systems located in the east and on the east side of the dish on systems located in west. If you are somewhere in the middle of the continent, observe how they are mounted on existing systems in your area or ask your satellite retailer which method is the correct one for your installation.

Refer to the Universal Azimuth-Elevation Look Angle previously presented to find the elevation angle for the satellite that is furthest to the west and furthest to the east of your location. Before tightening the clamp that goes around the arm, use the inclinometer to read the elevation angle of the dish. The arm is at its most retracted position when first installed. You will want the antenna pointed just below the lowest satellite at your location when the arm is in this position. Prop the antenna at this elevation angle, and then tighten the clamp on the arm just enough to hold it in place.

DISH ALIGNMENT SETTINGS

You have now completed the basic construction of the dish and are eagerly anticipating your first pictures from space. To fine tune those pictures and get your dish to track, you, or someone you can communicate with, will need to watch the TV screen while adjustments are being made to the antenna. You may even want to bring the TV and receiver out to the dish. If you have purchased a dual-band receiver, be sure that it is set to C-band before proceeding.

At this point, it is best to connect the satellite receiver directly to the TV and not through a VCR, video switcher, splitter, or any other device. The appropriate receiver output connector will be labeled "To TV" or "RF OUT." On most receivers, the output signal can be switched between two VHF channels (two UHF channels in some areas). Select a VHF (or UHF where applicable) channel which is not in use in your area. Tune the TV set to receive the selected TV channel.

It is also a good idea to follow the step-by-step procedures provided in the owner's manual. In most cases, the indoor unit also will prompt you with instructions that are displayed on the TV screen or on the receiver's front panel.

You first may be instructed to "Set East" and "Set West" limits. This is referring to the limits of travel for the actuator arm or horizon drive.

Although the motor has a slip clutch to prevent damage when the arm is extended or retracted completely, it is best to set the receiver's programmable limits at positions before the arm reaches these points. Many motors come with built-in limit switches that will shut off the motor at designated points.

Setting the east and west programmable limits. The receiver's programmable limits need to be set to stop the travel of the arm or horizon drive before the built-in limit switches are engaged or before the drive's physical limits are reached. If the arm does reach its full length or is retracted completely, the motor's slip clutch will start making loud "clicks." Stop immediately. If the arm becomes stuck in this position, take the motor off and insert the blade of a heavy screwdriver in the slot where the motor engages the arm. Turn the screwdriver just enough to loosen the arm, then put the motor back in place.

The idea here is to set the limits so they are just past the last satellite at either end of the satellite arc but before the mechanical limits of the drive. You can determine the elevation angles for the last satellite to the east and west by using the chart previously provided.

You may find that, after you have started to receive satellite pictures, you have not left yourself enough room to jog the antenna past the lower or upper satellite (thus making sure you have the best possible picture). In this case you may have to loosen the clamp on the arm and physically adjust it a little, after which you may have to reprogram the limits. An 18-inch arm should be the length required for any 10-foot

(3m) or smaller antenna and also most 12-foot (3.7m) antennas. Longer arms are available for larger dishes.

Check to see if you need to reset the limits before programming in the locations of the individual satellites. Otherwise, you may delete programming information that you entered into memory prior to resetting the limits.

Setting the polar axis elevation of the dish. Extend the drive until the dish is looking at the highest point in the sky (arc zenith). Set the inclinometer onto the polar axis of the dish. Adjust the polar axis of the mount to the correct elevation angle for your location.

Setting the declination. Declination is the offset angle between the polar axis of the mount and the rim of the dish that permits the antenna to precisely track the Clarke Orbit. The declination angle is a direct function of the latitude of the site location. The declination setting must be adjusted to the figure supplied by the manufacturer for your specific latitude. If you need to compute the declination, you can find out the approximate value by using the chart provided on page 119.

With a modified polar mount antenna, correct tracking of the total Clarke Orbit is only possible when the declination has been properly set. Set the inclinometer on the back plate of the antenna (same plane as the rim of the dish). Adjust the mount's declination to set the antenna for an elevation angle that is equal to the polar axis angle plus the number of degrees of declination for your area. The dish should be looking down slightly from the angle of the mount.

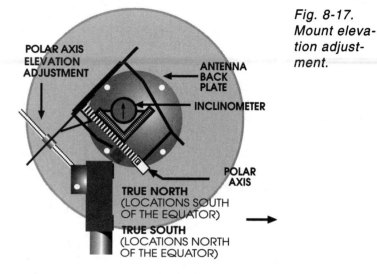

Fig. 8-17. Mount elevation adjustment.

POLAR AXIS ELEVATION ADJUSTMENT

ANTENNA BACK PLATE

INCLINOMETER

POLAR AXIS

TRUE NORTH (LOCATIONS SOUTH OF THE EQUATOR)

TRUE SOUTH (LOCATIONS NORTH OF THE EQUATOR)

TRACKING PROCEDURES

Now is the time to begin tracking the satellites and programming their positions into memory. This is not difficult now that you have everything set. When moving the dish to the east or west, the "look angle" of the dish should now follow a curve similar to that of the Clarke Orbit.

You may have a satellite receiver that was designed to automatically find the satellites and program their locations into memory. However, even these so-called "smart" receivers will require you to find and identify one upper and one lower satellite. There also are automatic dish aiming and polarization adjustment features on some receivers. If you choose to use these auto features, first follow the steps presented below to get your dish to track the Clarke Orbit. Once you are confident that you are receiving your best signal on a lower and upper satellite and have programmed them into the receiver, you are ready to turn on the auto program feature.

The accuracy of these auto features primarily depends on which brand of receiver or IRD you have

Fig. 8-18. Mount declination adjustment.

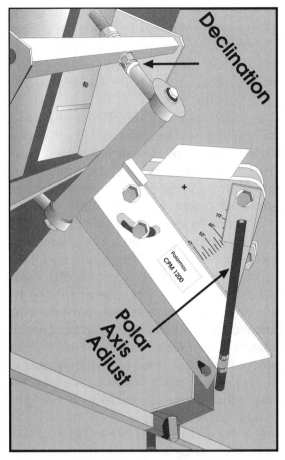

purchased. Some receivers occasionally miss a satellite or two. In this case you will have to manually peak the position of the dish and polarization of the feedhorn. At installation time, we recommend you turn these features off.

Some set-top boxes have a scanning feature that is handy for locating satellites. These receivers scan repeatedly through all of the available satellite channels at a rapid pace, providing you with glimpses of the active transponders. Channels will flash by on your TV screen as you move the dish past a satellite. You can go back to where the flashes occurred, turn off the scan feature, and identify the satellite by comparing the programming you encounter to a recent copy of one of the available satellite TV program guides. Some satellites only carry a few active TV transponders. Without the scan feature, you would need to select an active transponder and tune to the correct polarization before you could find the satellite.

Preview the dish set-up and programming sections in your receiver manual. Also read through these next steps to get familiar with the overall procedure you are about to perform. Finally, initiate each of the steps below.

Inclinometer readings for the satellite look angles (the elevations of the satellites available in your area) should be taken on the back plate of the antenna or on a surface parallel to the plane of the rim of the dish.

1.) Determine the elevation of the lowest satellite available from your location and move the dish until the inclinometer registers that elevation.

2.) Engage the set-top box's scan feature or set the unit to a channel that should have programming on it. A 24-hour service, even if it were scrambled, would be a good choice. Refer to a current satellite program guide to find a suitable selection.

3.) Push the dish to the right or left so that it rotates on the pole until you see a flash of video on your TV screen. Turn off the scan and go through the channels until you find video.

If the video seems to zip off the screen as soon as you stop on the channel, try changing the receiver's polarization format. That is, if you have your odd channels set for vertical polarization and your evens for horizontal, change the format to odds-horizontal, evens-vertical. You may not even know the present polarization format for your system. A change in format changes the relationship of the feedhorn pick-up probe by 90 degrees. If changing the format keeps the video on the screen, then it has to be the correct setting for that satellite on your system.

Select the best skew setting for the odd and even channels. The skew is the fine tuning of the feedhorn polarization. This is required because the satellite's signal polarization is only truly "horizontal" or "vertical" when the satellite is positioned along the same longitude as the installation site. The skew and format buttons are found either on the front of the receiver or on the remote control. Many receivers with on-screen graphics require you to choose these functions on the menu.

Most receivers have either LED signal level indicators on the front panel or digital level indicators presented by on-screen graphics. The level indicator provides a better reading of peak satellite performance in contrast to just viewing the picture on the TV screen.

4.) For locating satellites in the lower section of the Clarke Orbit, push the dish right and left on the pole and move the drive east and west in slight increments until you are satisfied you have the strongest signal. Then tighten the mount bolts onto the pole so the dish won't rotate. Do not tighten them down firmly yet, however. Note the numerical reading provided on the receiver front panel or on-screen display that corresponds to the dish position at this location. This is a relative number that changes as the dish moves.

5.) Move the dish east or west until you reach arc zenith for your location. Select an active transponder for the satellite closest to the east or

west of arc zenith or turn the scan control on. Now move the dish in the direction of the nearest satellite. If all the settings have been done correctly, you should not have a problem finding it.

6.) For locating satellites that are in the upper section of the Clarke Orbit, jog the actuator drive east and west and adjust the elevation bracket (on the mount of the dish) up and down slightly. You should not have to move the elevation setting very much because you already have pre-set it for your location. The rule of thumb here is rotate the dish on the pole and use the actuator to receive lower satellites, but do not rotate the dish on the pole to receive upper satellites. Adjust the elevation of the dish and use the actuator to receive upper satellites, but do not adjust the elevation to receive lower satellites.

7.) Fine tune your tracking by repeating these steps until the satellites in both the upper and lower sections of the Clarke Orbit appear at their maximum signal strength without requiring any further adjustments. If this cannot be done,

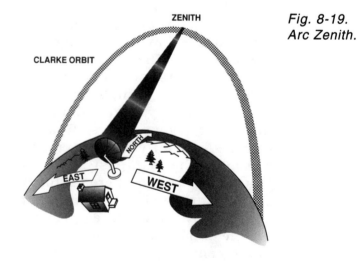

Fig. 8-19.
Arc Zenith.

Fig. 8-20.
Declination
offset angles
for sites in
North
America.

N. or S. LATITUDE	NORTHERN CITY/REGION	POLAR AXIS ANGLE	ARC ZENITH	DECLINATION ANGLE
65	Fairbanks	65.52	73.32	7.80
60		60.59	68.06	7.47
55		55.66	62.72	7.06
54		54.67	61.64	6.97
53		53.67	60.56	6.89
52		52.68	59.48	6.80
51		51.69	58.39	6.70
50	Winnepeg	50.69	57.31	6.62
49	Western Border U.S./Canada	49.70	56.21	6.51
47	Duluth	47.70	54.02	6.32
45	Minneapolis	45.71	51.82	6.11
43	Milwaukee	43.72	49.61	5.89
41	Salt Lake City	41.71	47.38	5.67
40	Philadelphia	40.71	46.27	5.56
39	Kansas City	39.70	45.15	5.45
37	Norfolk	37.69	42.90	5.21
35	Albuquerque	35.68	40.65	4.97
33	Dallas	33.67	38.38	4.71
31		31.64	36.11	4.47
30	New Orleans	30.63	34.96	4.33
29		29.62	33.82	4.20
28	Tampa	28.61	32.68	4.07
26	Miami	26.58	30.38	3.80
20	Kona, Hawaii	20.47	23.45	2.98
15		15.37	17.62	2.25
10	Costa Rica	10.26	11.77	1.51
05	Bogota	05.13	05.89	0.76

(ALL ANGLES ARE IN DEGREES)

you have probably made a miscalculation in the elevation or declination settings. Once you have the dish tracking properly, firmly snug the bolts that secure the mount onto the pipe. Placing a bolt all the way through the pipe may be desirable in high wind areas. Mark the pole and the mount for later reference.

IRD CONNECTIONS

The final installation step is to connect the IRD to your TV set. One method is to hook up a coaxial cable directly from the "To TV" ("RF Out") output of the receiver to the "Antenna" input of the TV set. Your local antenna wire, which formerly was connected directly to the TV set, should now be connected to the "An-tenna" input on the back of the IRD. To watch local channels, turn the satellite receiver off and the local channels will automatically appear on their respective channels on the TV.

If your satellite receiver does not have this feature, you will need to purchase an A/B switch from your local satellite or electronics store. Both the satellite receiver and local antenna connect to the "A" and "B" input ports on this switch, while the single output port connects to the antenna input on the TV. This switch will have to be changed manually whenever you want to go from one to the other. You also will need to use an external A/B switch if a second TV is connected to the satellite receiver and one person wants to watch local TV while the other is watching a satellite TV program.

There are other options when it comes to connecting the IRD to your TV set or home theater system. The video and stereo audio outputs on the back of the IRD can be connected directly to a TV monitor or to a VCR for recording programs. Whenever a VCR is part of the overall entertainment system, the output of the VCR must be connected to the TV set and the local antenna is usually also connected to the "Antenna" input of the VCR. The IRD's stereo audio outputs can be connected directly to the stereo inputs of the TV set or to the auxiliary inputs of your stereo system's audio amplifier or tuner.

INTERNET/DIGITAL DTH INSTALLATIONS

Home satellite receiving systems are now available that allow you to receive a digital DTH bouquet AND high-speed Internet access from a single dish. In 1997, Hughes Network Systems introduced a new version of their DirecPC satellite Internet system called the DirecDuo, which uses a 21-inch elliptical satellite dish to deliver Internet downloads to a home computer at speeds of up to 400 kb/s while also allowing other individuals in the home to simultaneously access programming from the DIRECTV and USSB DBS satellites.

The DirecDuo system uses a slightly larger dish with a different shape than what is normally required for a dedicated DBS installation. DirecDuo also requires the use of a special "tri-mode" feed with two LNBs that can simultaneously receive signals from two different satellites,

with one carrying the digital DTH channels and the other acting as your Internet Service Provider (ISP) in the sky. The computer requirements for a DirecDuo installation include a personal computer with Pentium Processor and the Windows 95 or Windows NT 4.0 operating system, 16 Mbytes of RAM, 20 Mbytes of free hard disk space and a modem. The DirecDuo installation kit comes with a 16-bit ISA/PCI card that plugs into an unused expansion slot inside your PC, a Windows 95 compatable software package, the installation and user manuals, a DIRECTV/USSB compatible digital IRD and three coaxial cables: two for connecting to the digital IRD and one for connecting the dish to ISA/PCI card that must be installed inside your personal computer.

Existing DIRECTV and USSB customers can upgrade their systems without having to purchase another digital IRD. Everything else mentioned in the DirecDuo installation kit will be required.

Most of the dish installation procedures previously discussed for small dish application—such as determining line-of-site access to the satellites, adjusting the dish to the correct coordinates, running the cables and grounding the system—equally apply to the installation of the DirecDuo system. There are some major differences, however, between the installation of a digital IRD and a DirecDuo set-up. All digital IRDs come with factory installed software so that they can go right to work as soon as you take them out of the box and hook them up. With DirecDuo, you must install and configure both the ISA/PCI card and the software package so that it can function cor-

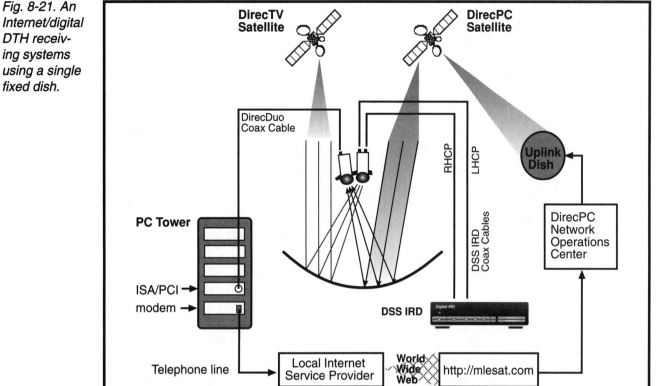

Fig. 8-21. An Internet/digital DTH receiving systems using a single fixed dish.

rectly with the computer hardware and software that you already have. There are also differences in how you align the dish for peak performance. With a digital DTH system, you can use the IRD's built-in signal meter and audio tone generator to find the antenna's sweet spot. With DirecDuo, your PC and the DirecDuo computer software provide the indicators that will notify you when the dish has been correctly aligned to pick up the ISP in the sky.

After the system has been installed, you will continue to use a modem and a local ISP to initiate all your Internet download requests. However, your data requests will now contain routing codes that will instruct any web site that you access to forward the data directly to the DirecPC Network Operations Center (NOC), which then uplinks the In-

ternet data to their satellite. Like an IRD, your ISA/PCI card has an "address code" that the NOC uses to forward the data on to your system. This address code also provides your computer with the ability to distinguish the data that is intended for your terminal from data that is intended for other DirecDuo customers. As is the case with all subscription services, you will need to pay a monthly fee to access the system. The DirecDuo system also gives the NOC the ability to switch off your ISA/PCI card in the event that you fail to make payment.

SIGNAL DISTRIBUTION IN THE HOME
The simplest signal distribution system consists of several runs of 75-ohm coaxial cable and a cable TV accessory called a "signal splitter." Splitters come in different configurations: two-way, three-way, four-

way, and so on. A short length of RG-59 cable, called a coaxial jumper, connects the RF modulator output of the IRD to the input port of the splitter. One output port of the splitter supplies the signal to the primary TV set. The other splitter output port or ports will send the signal to a TV set in the master bedroom or to any other TV set locations throughout the home. Extra splitters can be added down the line at any time to accommodate additional TV sets.

Whenever the TV signal is divided, the strength of the signal at any subsequent locations will be reduced. The RF modulator contained in most set-top boxes should be capable of supplying an adequate signal level to two or more TV sets. Larger distribution systems, which may have long runs of cable, may require the insertion of a "line amplifier" at some point midway between the IRD and those TV sets which are located at the far end of the line. This will boost the level of the RF signal so that all TV sets connected to the distribution system can display noise-free pictures.

Households equipped with multiple TV sets often have one master off-air antenna that is connected to the TV sets scattered throughout the home. A splitter, which is usually installed in the attic, divides the off-air antenna signal and sends it on to each TV set. If this is the case at your home, a device called a "combiner" should be inserted in the cable line at some point in front of the signal splitter. The combiner acts like splitter in reverse. It has two input ports that receive signals from the off-air antenna and the IRD's RF modulator. The output of the combiner produces a broadband signal that contains both the off-air TV channels as well as the satellite TV channel.

Although it would be easy enough to just use a splitter in reverse to combine the two signal sources, this is not a good idea. The signal output of the combiner will be much "cleaner" than the output of a splitter used in a reverse mode. The single output of the combiner, however, can be connected to one or more splitters to distribute the satellite signal and off-air channels to every TV set connected to the system.

Each TV set in the home can switch to the TV channel that corresponds to the IRD's RF modulator. There usually is a switch on the back of the IRD that gives you the option of selecting one of two available stan-

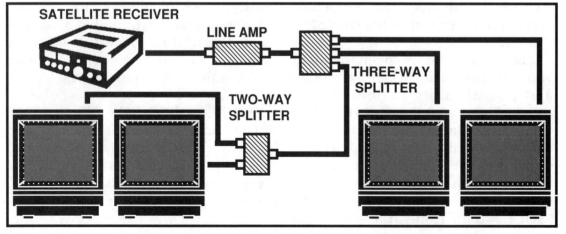

Fig. 8-22. In-house multiple TV set wiring diagram.

Fig. 8-23 &
Fig. 8-24.
Distribution
systems for
single polar-
ization, two
IRDs; and dual
polarization,
multiple IRDs.

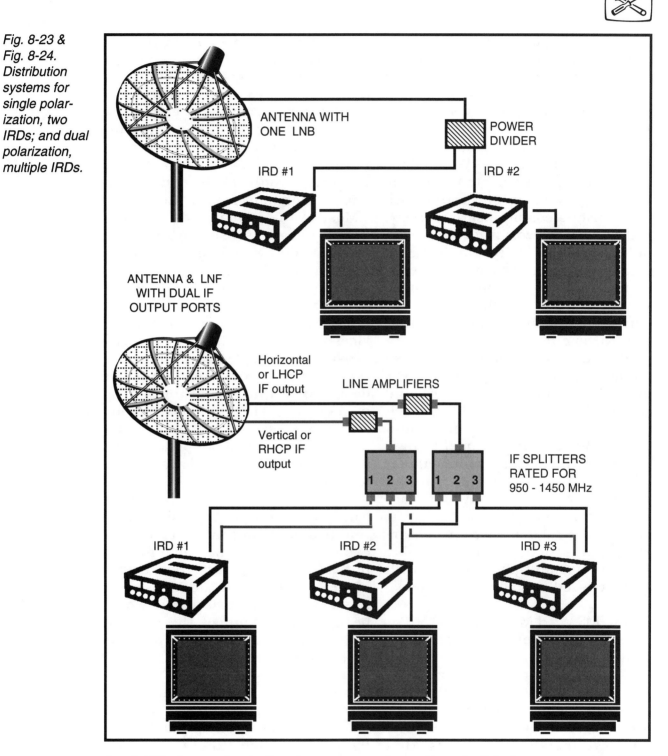

dard TV channels to be the output channel of the RF modulator.

An alternative method is to use a low-power UHF TV transmitter to send TV programming from the living room to the bedroom without having to drill holes or string wires. UHF (short for ultra high frequency) signals can pass through walls and floors, usually up to a distance of about 200 feet. This device, which connects to the IRD's video and audio outputs, is usually placed on top of the TV set or in some other high, unobstructed location.

Some models even transmit UHF frequencies that can be received by the UHF tuner on your TV set. Others use a special remote receiver that connects to the TV set's antenna input. The remote receiver is tuned to pick up special frequencies that various communications regulatory bodies worldwide have set aside for home video applications. Additional receivers can be ordered so that multiple TV sets throughout the home can simultaneously receive the satellite TV program that the IRD is currently receiving.

Your next-door neighbors also could purchase one or more of these special receivers, and they too could watch what you are watching. Such an arrangement, however, would inevitably lead to arguments over program choices. Reaching agreement about what to view in your own home is tough enough without dragging your neighbors into the combat zone. Other signal distribution arrangements that are more suitable for serving multiple dwellings will be discussed in the next chapter.

SYSTEMS WITH MULTIPLE IRDS

The next level of independence requires two or more set-top boxes, with each IRD independently choosing from the TV services that are available from a single satellite or group of collocated satellites. The simplest installation of this kind would limit the reception of all IRDs to satellite TV signals that share the same sense of polarization. The LNF generates an output signal that contains all incoming satellite signals of a given polarization. The LNF out-

put can be divided so that two or more set-top boxes can simultaneously access this broadband signal and independently choose any available program service that the signal contains. The major limiting factor with this approach, however, is that it restricts all set-top boxes to receiving only those services that share a common polarization.

Most of the available digital DTH delivery systems use two senses of polarization, which may be right-hand and left-hand circular (DBS satellites) or vertical and horizontal (Ku-band digital DTH). This effectively doubles the number of TV services that the satellite operator can provide. The simple distribution system previously described could be designed so that one of the set-top boxes could switch LNF polarization from one sense to the other. Remember that any change in polarization would affect the channel availability for every set-top box in the home.

A best solution would be to install a dual-polarization "orthomode" LNF that has two IF output ports, one for each sense of polarization (or alternatively a dual-polarization orthomode feedhorn and two LNBs). For this system, two coaxial cables would be needed to connect the outdoor unit to the various set-top boxes installed in the home.

Other accessories—such as line amplifiers and IF signal splitters — also will be required (see Fig. 8-24 for an example of one possible configuration). Each IRD in your system also would need to have dual IF input ports on their rear panels, which indicates the presence of a built-in polarization switch. This vastly simplifies the installation a multiple IRD distribution system.

MULTI-DWELLING SIGNAL DISTRIBUTION SYSTEMS

Cost-effective coverage of multiple dwellings requires the construction of either a satellite master antenna TV (SMATV) system or a Multi-channel Multi-point Distribution System (MMDS). In condominiums, caravan parks, and public buildings such as hotels, motels, and health care facilities, SMATV systems can be very practical. The owners or managers often install these systems as a commercial venture. Tenant co-operatives and neighborhood associations also may become involved in the installation of these systems. MMDS systems are the more practical solutions for servicing relatively wide coverage areas. In the vast majority of cases, both of these distribution systems rely on satellites to provide the lion's share of the programming that the system operator will offer to its subscriber base.

SMATV OVERVIEW

SMATV systems are mini cable TV systems that select a few services off the satellites, change them to regular TV channels, and send them down a cable to the individual residences. The traditional approach is to use multiple satellite receivers or IRDs, with each unit dedicated to the reception of a single service. The video and audio from each receiver is connected to an external RF modulator. The output of the RF modulator corresponds to an unused or open channel in the SMATV system. Some manufacturers provide commercial rack mounted receivers and modulators ex-

pressly for this purpose.

The following chapter presents a general overview of the design and installation of private cable (called SMATV) distribution systems. It is the intent of this chapter to inform newcomers to the field about the basic technical requirements involved.

Our tour of the various stages of a SMATV system will begin with the antenna farm and the "head end" which contains satellite receivers and off-air amplifiers, as well as the signal modulators and combiners. The combined throughput of all signals is then fed into a cable distribution system which supplies programming to multiple locations in a building or to multiple buildings within a given subscriber coverage area. Subsequent sections covering SMATV will explain how to design a distribution system as well as how to select and install the various essential components. Information on MMDS systems appears toward the end of this chapter.

THE ANTENNA FARM

Most of the requirements concerning satellite antennas that have previously been discussed in this book also pertain to the antennas used by any multi-dwelling signal distribution system. It should be pointed out, however, that the size of the satellite antenna or antennas used in an SMATV system are invariably larger than what are used for consumer applications. The idea is to have a very good signal arriving at the head end so that any degradations

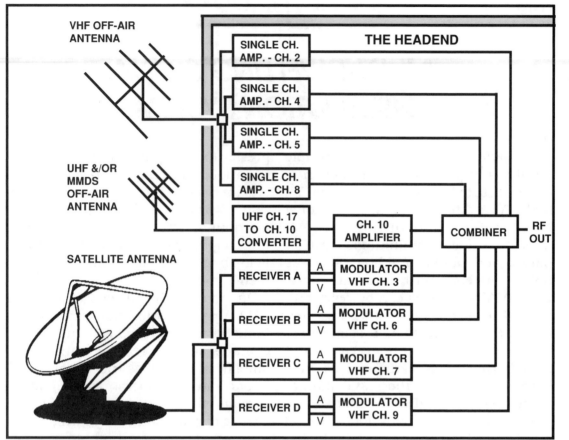

VHF OFF-AIR ANTENNA

UHF &/OR MMDS OFF-AIR ANTENNA

SATELLITE ANTENNA

THE HEADEND

SINGLE CH. AMP. - CH. 2

SINGLE CH. AMP. - CH. 4

SINGLE CH. AMP. - CH. 5

SINGLE CH. AMP. - CH. 8

UHF CH. 17 TO CH. 10 CONVERTER

CH. 10 AMPLIFIER

RECEIVER A — MODULATOR VHF CH. 3

RECEIVER B — MODULATOR VHF CH. 6

RECEIVER C — MODULATOR VHF CH. 7

RECEIVER D — MODULATOR VHF CH. 9

COMBINER

RF OUT

Fig. 9-1. A typical head end providing satellite, VHF and UHF channels.

caused by processing and distribution will not greatly affect the quality of the signal arriving at any single dwelling. For example, any noise in the video or audio can cause problems with decoder signal authorization.

With programming spread out over so many different satellites, it is almost a given that you will need to be able to receive signals from several different satellites to get access to all the channels. Sometimes an option to consider is the use of a "multi-feed"—a special antenna feed support system that allows you to mount multiple feed horns for the reception of several satellites simultaneously from one dish. However, only one satellite can be received at boresight, where the highest level of signal con-

centration occurs.

Any shallow dish can potentially generate multiple focal points, with each secondary focal point receiving signals from angles that are offset from the parabolic antenna's axis of symmetry. Each of these secondary focal points, however, receive signals that are reflected from just a portion of the total

Fig. 9-2. Satellite industry pioneer Bob Cooper maintains this cable head end, which provides service to several small villages in Far North, New Zealand.

Fig. 9-3. A shallow parabolic dish equipped with a multi-feed/LNB assembly, also known as a "cross-eyed" feed installation.

permit, it is almost always better to install another dish to receive programming from an additional satellite.

Mass production has brought the cost of consumer grade dishes down quite a bit, tempting many commercial installers to use substandard antennas. Commercial grade antennas are more rugged, designed to withstand high winds and inclement weather without a disruption in reception. This is primarily a function of the mount, or supporting structure. If you grab the edge of the dish and pull in either direction, a commercial antenna will show a lot of resistance, barely moving.

surface area of the antenna's reflector. The additional feedhorns also create a larger blockage zone than the single feed system, which acts to reduce the efficiency of the dish. To make up for these deficiencies, any antenna that is equipped with a multifeed assembly will need to be larger than a parabolic dish that is equipped with just a single feed. In the long run, if space and finances

HIGH TIME AT THE HEAD END

The SMATV system's "head end" serves as the central receiving center for all of the signals being imported for distribution to subscribers. One or more satellite dishes, as well as multiple off-air VHF and UHF antennas, are used to intercept the available programming resources and route their signals into the head end. Designing the head end really begins outside at the antenna farm. By determining your programming package, you can set about the task of selecting the number and types of antennas (both satellite and off-air) needed to accomplish your goal.

MULTIPLE SATELLITE IRDS

The number of IRDs that you can have in your head end is virtually unlimited. After all, the design of your system can include multiple dishes, with each antenna pointed at a different satellite

Fig. 9-4. The relatively "shallow" parabolic antenna (f/D from .35 to .45) creates multiple focal points that are offset at angles away from the antenna's axis of symmetry.

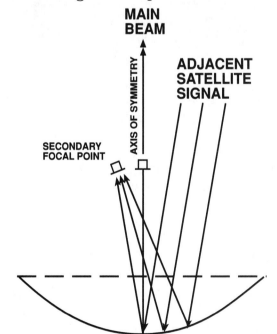

MAIN BEAM

ADJACENT SATELLITE SIGNAL

AXIS OF SYMMETRY

SECONDARY FOCAL POINT

carrying dozens of TV channels, as well as equipment racks loaded with commercial satellite receivers, with each receiver tuned to a different satellite service.

Signals from each LNB in the system are split as many times as necessary to accommodate the number of receivers picking up channels from a like-polarized set of transponders on a satellite. Off-air signals received by a terrestrial TV antenna are run either to a broadband amplifier and processor or to multiple single-channel amplifiers and processors.

Video and audio from each satellite receiver is fed directly to a modulator. In the case of encrypted satellite services, if the receiver is an integrated receiver/decoder (IRD) the encrypted video and audio information is decoded and then sent to the modulator, again leaving the IRD as audio and video. If a separate decoder is used, a composite signal from the receiver containing both the encrypted audio and video is fed to the decoder. The decoder unscrambles the audio and video information and sends it on to the modulator as separate signals.

THE OFF-AIR MASTER ANTENNA

Nearly all SMATV installations will pull some of their channels from local off-air broadcast networks. Unlike satellite reception, which can be depended on to supply perfect audio and video from almost any location with a clear view of the sky, off-air systems can be plagued by a number of problems. Terrestrial broadcast signals travel more or less by line of sight. Obstructions such as tall buildings, hills or mountains can actually block the incoming TV signal, reducing signal strength to little or nothing, even though the transmitting tower may be relatively close by. These ob-

Fig. 9-5. New modular approaches to the design of SMATV head ends provide operators with the opportunity to purchase complete turnkey installations. As many as 24 signal processing modules can be installed in this cabinet. (Photo courtesy of Hirschmann.)

structions also can play havoc with the signal by bouncing and reflecting the signal, creating multiple, out of phase versions of the same program service that arrive at the antenna out of synchronization by several milliseconds. This produces trailing images, commonly called "ghosts."

SELECTING AN OFF-AIR ANTENNA

There are two basic types of off-air master antenna television (MATV) aerials available, regardless of whether they are for VHF or UHF TV reception. These

Fig. 9-6. An inside view of a head end equipped with rack-mounted receivers, decoders, and RF modulators.

are known as Broadband and Single Channel antennas. Broadband TV antennas are designed to receive all of the signals occupying the entire VHF (Channels 2 - 13 in the U.S.) or UHF (Channels 14 and above) frequency spectrums. Single channel antennas are more selective, cut exactly to measurements that will allow them to resonate strongly with a specific channel, rejecting all others. You must select the type of off-air antenna that will work best at each particular location.

Most broadcast networks transmit within major metropolitan areas. If the head end site is thirty to forty miles to the south of the city, then all terrestrial TV signals will come from a single direction. A broadband antenna pointed to the north will receive all the available channels simultaneously. This broadband signal later can be divided and fed to amplifiers for each individual TV channel.

When installing a system inside city limits, you may find that one TV transmitter is located north of the site location, another to the east, and yet another to the south. In this case, you will need a dedicated, single channel antenna that has been specifically designed to receive each local TV channel

frequency. Each of these separate antennas also must be pointed in the direction of the broadcaster's transmission tower.

THE PRE-AMP

The further the site is away from the broadcast channel's transmitter, the weaker the signal will be. If the site is located in a fringe area that receives relatively weak signals, a small amplifier, called a pre-amp, should be installed at the antenna to boost the incoming signal to the level that is required by the main amplifier in the head end for that channel.

THE BROADBAND AMPLIFIER

The off-air antenna signals are fed to a single broadband amplifier located inside the head end. If you are using a broadband antenna, then a single coaxial cable connects the antenna to the broadband amplifier that feeds your distribution system. Whenever the MATV system has multiple antennas, the signals from each antenna must be combined together onto one cable before arriving at the input of the amplifier.

SINGLE CHANNEL AMPLIFIERS

A preferred installation method is to supply a dedicated amplifier for each terrestrial TV channel to be carried by the system. Each amplifier is tuned to a specific channel. The system designer, therefore, must know at the outset which channels will be received by the MATV antenna system.

Whenever the MATV system uses multiple, single-channel antennas, the coaxial cable from each antenna is fed to its corresponding amplifier. A system with a single broadband antenna has a single coax going to the head end. This cable must pass through a splitter, which divides the signal for delivery to each amplifier.

Fig. 9-7. This bandpass filter surpresses out-of-channel modulator emissions with an insertion loss of less than 1.5 dB. (Courtesy Communications & Energy Corp.)

BANDPASS FILTERS

The ultimate goal is for each dedicated amplifier to only boost the signal of one TV channel. Bandpass filters are used to eliminate other irrelevant channels that may otherwise be received and sent down the cable from the MATV antenna to the amplifier. The bandpass filter strips away all irrelevant signals, passing only the one desired channel. Bandpass filters are passive in that they require no electrical hook-up or provide any amplification.

CHANNEL PROCESSORS

Processors differ from ordinary amplifiers in that they convert the terrestrial TV channel received from a local broadcaster to another TV channel frequency that has no local off-air channels competing with it. Problems often result whenever off-air channels are sent through the cable distribution system on the same channel frequencies used to transmit the signals through the air. Processors are used to prevent the phasing problems which cause the appearance of ghosts in the TV picture. The processor places the TV signal on an entirely new channel so there is no possibility of multiple versions of the same signal arriving out of phase at the TV set. The main disadvantage to this setup is that the local broadcast networks will appear on channels that differ from the listing in the local newspaper or TV publication. This can be confusing for the new system viewer.

OFF-AIR UHF TO VHF CONVERSION

Most major U.S. cities have at least one UHF TV channel. UHF frequencies have greater loss when passing through a given distance of coaxial cable than do VHF signals. Most private cable systems convert the UHF channel to a lower frequency channel, generally an unused VHF channel. The complete installation begins with a broadband or a custom-tuned UHF antenna and UHF pre-amp if the signal needs boosting. The output signal is fed to the UHF converter which strips the video and audio information off of the original UHF channel frequency and places this information onto a specific VHF channel frequency. This converted output is sent to a single channel amplifier and then combined with all of the other channels in the head end.

THE RF MODULATOR

The decoded audio and video information leaving the satellite receiver, decoder or IRD is sent onward to the RF modulator. Each modulator has its own RF output, tuned to one specific VHF, Midband, Superband, or in some cases, UHF TV channel frequency.

Fig. 9-8. Trunk Line, VHF, Alpha, Midband, Superband, Ultra Band, and UHF Cable and Broadcast TV Channel Assignments for North America (Frequency in MHz).

Channel Designation	Ch. No.	Visual/Audio Frequencies
BACK FEED TRUNK LINE CHANNELS		
T-7	—	7.00/11.50
T-8	—	13.00/17.50
T-9	—	19.00/23.50
T-10	—	25.00/29.50
T-11	—	31.00/35.50
T-12	—	37.00/41.50
T-13	—	43.00/47.50
VHF TV CHANNELS - LOW BAND		
2	2	55.25/59.75
3	3	61.25/65.75
4	4	67.25/71.75
5	5	77.25/81.75
6	6	83.25/87.75
ALPHA CABLE TV CHANNELS		
A-5	95	91.25/95.75
A-4	96	97.25/101.75
A-3	97	103.25/107.75
A-2	98	109.25/113.75
A-1	99	115.25/119.75
VHF TV CHANNELS - HIGH BAND		
7	7	175.25/179.75
8	8	181.25/185.75
9	9	187.25/191.75
10	10	193.25/197.75
11	11	199.29/203.75
12	12	205.25/209.75
13	13	211.25/215.75
MIDBAND CABLE TV CHANNELS		
A	14	121.25/125.75
B	15	127.25/131.75
C	16	133.25/137.75
D	17	139.25/143.75
E	18	145.25/149.75
F	19	151.25/155.75
G	20	157.25/161.75
H	21	163.25/167.75
I	22	169.25/173.75

Channel Designation	Ch. No.	Visual/Audio Frequencies
SUPERBAND CABLE TV CHANNELS		
J	23	217.25/221.75
K	24	223.25/227.75
L	25	229.25/233.75
M	26	235.25/239.75
N	27	241.25/245.75
O	28	247.25/251.75
P	29	253.25/257.75
Q	30	259.25/263.75
R	31	265.25/269.75
S	32	271.25/275.75
T	33	277.25/281.75
U	34	283.25/287.75
V	35	289.25/293.75
W	36	295.25/299.75
ULTRABAND CABLE TV CHANNELS		
AA	37	301.25/305.75
BB	38	307.25/311.75
CC	39	313.25/317.75
DD	40	319.25/323.75
EE	41	325.25/329.75
FF	42	331.25/335.75
GG	43	337.25/341.75
HH	44	343.25/347.75
II	45	349.25/353.75
JJ	46	355.25/359.75
KK	47	361.25/365.75
LL	48	367.25/371.75
MM	49	373.25/377.75
NN	50	379.25/383.75
OO	51	385.25/389.75
PP	52	391.25/395.75
QQ	53	397.25/401.75
RR	54	403.25/407.75
SS	55	409.25/433.75
TT	56	415.25/419.75
UU	57	421.25/425.75
VV	58	427.25/431.75
WW	59	433.25/437.75
XX	60	439.25/443.75
YY	61	445.25/449.75
ZZ	62	451.25/455.75
AAA	63	457.25/461.75
BBB	64	463.25/467.75
CCC	65	469.25/473.75
DDD	66	475.25/479.75

Channel Designation	Ch. No.	Visual/Audio Frequencies	Channel Designation	Ch. No.	Visual/Audio Frequencies
EEE	67	481.25/485.75	—	32	579.25/583.75
FFF	68	487.25/491.75	—	33	585.25/589.75
GGG	69	493.25/497.75	—	34	591.25/595.75
HHH	70	499.25/503.75	—	35	597.25/601.75
III	71	505.25/509.75	—	36	603.25/607.75
JJJ	72	511.25/515.75	—	37	609.25/613.75
KKK	73	517.25/521.75	—	38	615.25/619.75
LLL	74	523.25/527.75	—	39	621.25/625.75
MMM	75	529.25/533.75	—	40	627.25/631.75
NNN	76	535.25/539.75	—	41	633.25/637.75
OOO	77	541.25/545.75	—	42	639.25/643.75
PPP	78	547.25/551.75	—	43	645.25/649.75
QQQ	79	553.25/557.75	—	44	651.25/655.75
RRR	80	559.25/563.75	—	45	657.25/661.75
SSS	81	565.25/569.75	—	46	663.25/667.75
TTT	82	571.25/575.75	—	47	669.25/673.75
UUU	83	577.25/581.75	—	48	675.25/679.75
VVV	84	583.25/587.75	—	49	681.25/685.75
WWW	85	589.25/593.75	—	50	687.25/691.75
XXX	86	595.25/599.75	—	51	693.25/697.75
YYY	87	601.25/605.75			
ZZZ	88	607.25/611.75	—	52	699.25/703.75
AAAA	89	613.25/617.75	—	53	705.25/709.75
BBBB	90	619.25/623.75	—	54	711.25/715.75
CCCC	91	625.25/629.75	—	55	717.25/721.75
DDDD	92	631.25/635.75	—	56	723.25/727.75
EEEE	93	637.25/641.75	—	57	729.25/733.75
FFFF	94	643.25/647.75	—	58	735.25/737.75
			—	59	741.25/745.75
UHF TV CHANNELS			—	60	747.24/751.75
			—	61	753.25/757.75
—	14	471.25/475.75	—	62	759.25/763.75
—	15	477.25/481.75	—	63	765.25/769.75
—	16	483.25/487.75	—	64	771.25/775.75
—	17	489.25/493.75	—	65	777.25/781.75
—	18	495.25/499.75	—	66	783.25/787.75
—	19	501.25/505.75	—	67	789.25/793.75
—	20	507.25/511.75	—	68	795.25/799.75
—	21	513.25/517.75	—	69	801.25/805.75
—	22	519.25/523.75			
—	23	525.25/529.75			
—	24	531.25/535.75			
—	25	537.25/541.75			
—	26	543.25/547.75			
—	27	549.25/553.75			
—	28	555.25/559.75			
—	29	561.25/565.75			
—	30	567.25/571.75			
—	31	573.25/577.75			

Fig. 9-8 (continued). Trunk Line, VHF, Alpha, Midband, Superband, Ultra Mand, and UHF Cable and Broadcast TV Channel Assignments for North America (Frequency in MHz).

UHF TV channels 52 through 69 will be phased out of broadcast service by the end of the FCC mandated simulcast period for the introduction of the new DTV standard. These frequencies will, however, remain available for cable and SMATV use.

The output signals for all of these modulators must be combined together into one single broadband "multiplexed" signal that is then distributed to each of the subscribers connected to the private cable system.

MODULATOR CHOICES

One of the best cable innovations in recent years has been the introduction of frequency-agile modulators. Even though most modulators remain dedicated to one single RF frequency for their entire service lives, frequency-agile modulators offer several advantages over fixed-frequency units.

Fixed-frequency or frequency-agile refers to the modulator's tuning. Fixed-frequency modulators come pre-tuned to one channel in the VHF, UHF, Mid-band, or other bands used in cable distribution. Frequency-agile modulators can be tuned and/or switched through the entire range of channels.

The electronics in a fixed-frequency modulator are less complicated than in those providing multi-channel selection and therefore are usually less expensive. For flexibility, the head end can be designed with a mix, with fixed-frequency units in primary use and one or more fully agile units available as backups for the system. A system like this holds the cost down and gives the operator fully redundant capability for all channels.

The first measurement usually listed in modulator specifications is the output level, measured in decibels, relative to one milli-volt (dBmV). Medium power commercial modulators have a per carrier output of 25 to 35 dBmV. High power models offer even more. Obviously, the higher the output, the more TVs it can supply with ample signal, or the longer distance it can travel down a cable before the signal needs boosting.

In contrast, the channel 3/4 modulator usually found in consumer satellite receivers and videocassette recorders generally has an output of about 5 dBmV. That's considered about the ideal amount needed for one TV, although it is enough to permit splitting the signal to feed additional TVs in a household, and still have an acceptable picture on each set as long as the cable runs aren't too long.

The output of the modulator is critical when designing a commercial head end. If you have a twenty-channel system with all modulators supplying an output of 55 dBmV, you cannot replace a malfunctioning unit with a modulator output of 35 dBmV. Since all channels on the TV must appear equally strong and clear, their signal levels must be equalized before leaving the head end.

If the system uses 35 dBmV output modulators, a higher-powered modulator, such as one with 55 dBmV, cannot be used as a substitute unless its output power level is reduced. Attenuator pads can be installed in series with the output of any high-power modulator. These serve to limit or drop the power down to the desired level without introducing impairments. Pads come in a variety of values (3, 6, 10, 20 dB, etc.). They can be mixed and matched until just the right level is found, keeping the outputs of all equipment evenly balanced.

MODULATOR FILTERING

Filtering in the modulator is as critical as setting the power level. For example, if the modulator's output is channel 3, there should be no channel 3 interference, or "artifacts," visible on either channel 2 or 4.

One part of the modulator is a single channel amplifier. Amplifiers often produce harmonics and spurious frequencies if the gain is set too high. These can show up as bleed-over or "ghost images" on adjacent channels. Harmonics can be a multiple of another frequency and the interference can show up several channels away. An ideal situation would be to leave a blank channel on either side of an active channel to avoid adjacent channel interference. That luxury is most often eliminated by the number of channels to be transmitted and the need to stay away from the higher frequency channels, which are subject to excessive loss over long distances.

External bandpass filters on the modulator output of high gain amplifiers have limited effectiveness. The better modulators made today contain an internal SAW (Surface Acoustic Wave) filter. They do a good job of keeping the RF within the appropriate channel bandwidth and are all but essential for installations where adjacent channel interference is a potential problem to be overcome. The key to success here is to keep power levels balanced. If you have a modulator with an output of 55 dBmV on channel 3, and only 35 dBmV on channel 2, signal bleed-over on channel 2 is inevitable.

Inordinately high audio levels can also cause interference. The audio subcarrier frequency is located near the lower edge of the video waveform envelope. If the audio level is too high, it can bleed onto the next channel down as well as interfere with its own video carrier. Backing down the audio gain will usually correct this kind of problem. To ensure that all audio subcarriers are at the same level, a local off-air station can be used for reference. The technician also can use a dB meter to measure the intensity of the audio subcarrier.

Interference problems also will occur whenever the channel's video level is set too high. Not only will white and other bright portions of the picture appear too "hot," but it also can cause the audio to "buzz." Most modulators have manual adjustments on the front panel for the video/audio input and output levels.

COMBINING OFF-AIR AND SATELLITE TV SIGNALS

There are several ways to combine signals. Each method has its own advantages and disadvantages. A motel with four off-air stations, three of the four major networks and PBS, for example, may want to add one satellite channel. In many motel systems you will find these four signals come from one MATV antenna which feeds into a single broadband amplifier.

Let's assume that all of the off-air stations are located in the VHF band on channels 2, 4, 5 and 8. This leaves channels 3, 6, 7, 9, 10, 11, 12 and 13 as open slots. To reduce the possibility of adjacent channel interference, try to keep an empty channel between the satellite modulator's output channel and any of the local stations. That eliminates channels 3, 6, 7 and 9 but any one of the four remaining channels (10, 11, 12 and 13) should offer sufficient insurance against adjacent channel interference.

Where is the best place to combine the signals? The answer will depend upon the existing setup, your new equipment, and a little trial and error. However, the final result must be equal signal strength on all channels. Combining channels of different strengths will result in cross channel interference and/or poor quality pictures.

There are two ways to achieve an even balance. One is to mix the signals before the broadband amplifier, then feed the combined signals into the amplifier boosting everything up to a common power level. If the satellite modulator produces an output signal that is stronger than the off-air signals an attenuator can be used to reduce its power (measured in dBs) to that of the off-air channels. However, you may obtain a closer match by sending the off-air stations through the broadband amplifier, and then combining the amplifier output and satellite modulator outputs. As long as both outputs are kept at about the same signal strength, either method will work adequately.

If we are only adding one channel and there is an open channel slot between it and any other active channel, we can use an inexpensive combiner. In a case such as this, the most simple and cost effective solution would be a two-way splitter. A common splitter can be installed in reverse to mix signals together. Some units are marked as splitter/combiners. The satellite modulator output is connected to one splitter/combiner output port and the off-air antenna signals to the other. The input port now becomes a combined output with all signals present together on one cable.

What if there are two, three, four or more satellite signals to mix with the off-air? Can a splitter still be used? That depends on how broad or how clean the modulator outputs and the off-air signals are. A quality signal stays within the confines of its assigned channel's bandwidth. A broad signal goes beyond these limits and interferes with channels on either side or more.

Using the channel lineup previously mentioned, two signals can be added on channels 10 and 13, leaving plenty of space between stations and little likelihood of adjacent channel interference. However, as more channels are added and adjacent slots are filled, another device called a combiner should be used.

The difference between a combiner and a simple splitter/combiner is the level of filtering. Combiners are designed specifically for mixing channels and have a built-in level of separation. They also are built to take in more signals, anywhere from 6 to 12 inputs, and multi-

plex them into one output.

Even combiners have their limits. A wide and dirty signal can still appear elsewhere up and/or down the band. If a particular modulator or strip amplifier is too broad, no amount of filtering can rectify the situation.

On most combiners, all inputs are the same. There are some brands and models where the input is labeled for a specific channel or pair of channels. These units have tuning slugs that allow you to adjust the input for one channel to improve the degree of isolation between stations.

ACTIVE VERSUS PASSIVE COMBINERS

Whenever signals are passed through a combiner, they will experience some loss in output, often in the neighborhood of 10 dB, sometimes even more. Passive combiners merely take multiple signals in and send them back out through the single output. When designing a system, you must plan for this loss, either by accepting the output as it stands, or by bringing it back up with additional amplification.

A few manufacturers have designed their combiners to address this loss by including an internal broadband amplifier. These are called "active" combiners. They depend on an AC source to power the broadband amplifier, restoring the signals to their original levels. Most active combiners do not incorporate any tuning separation between high-band and low-band channels. Due to its limited power output, it often is necessary to use another amplifier before the combined signal enters the distribution system.

USING MULTIPLE COMBINERS

If more channels are required than the combiner has inputs, two combiners can be used in tandem. The outputs from two combiners often can be mixed together using the humble two-way splitter discussed earlier. Some head ends even put the output of one combiner into the input of another and treat it as just another channel. Since these channels are going through two combiners, the output power levels on the individual modulators and strip amplifiers must be set correspondingly higher, to compensate for the extra loss.

The output from a combiner is too strong to connect directly to a TV. Many combiners have an attenuated tap off specifically for connection to a monitor.

LOOP-THROUGH COMBINING

In some cases, you may be able to combine channels using a method called a "loop-through." With this method, the output of one off-air strip amplifier is fed to an input (labeled "loop-through") on another strip amplifier. The output of the second strip amplifier now contains both signals, which can be fed into the next unit.

The loop-through will work fine if all signals are clean and power output levels are fairly equal. However, when there is a large variation between channels, or if extra filtering needs to be added due to adjacent channel interference, a regular combiner will be a better choice.

A QUESTION OF BALANCE

No matter which combination you choose, the quality of the output is the bottom line. Balancing each channel for the proper amount of power output is essential to maintain distortion and in-terference-free video. Have a monitor present at the head end, so you can immediately see the results of your adjustments. Any amount of interference at the head end will be multiplexed every time it is amplified in the distribution system.

In the end, it is the planning of these channel combinations that can make the difference between poor, average or good performance in a private cable system. The art of balancing channels together into a common output devoid of cross channel interference is the magic that defines the skilled technician.

Set the power output on all of your modulators and strip amplifiers so that they are more or less equal. For this task, you'll need a signal strength meter which measures in dBs. In the initial planning and design stage, you should have determined the dB level needed from the head end. The amount of power required can vary. The variance will depend on whether you have a main power distribution amplifier right after the head end (such as for a larger private cable distribution system) or whether the modulators themselves will supply all or most distribution power (such as for a hotel/motel system).

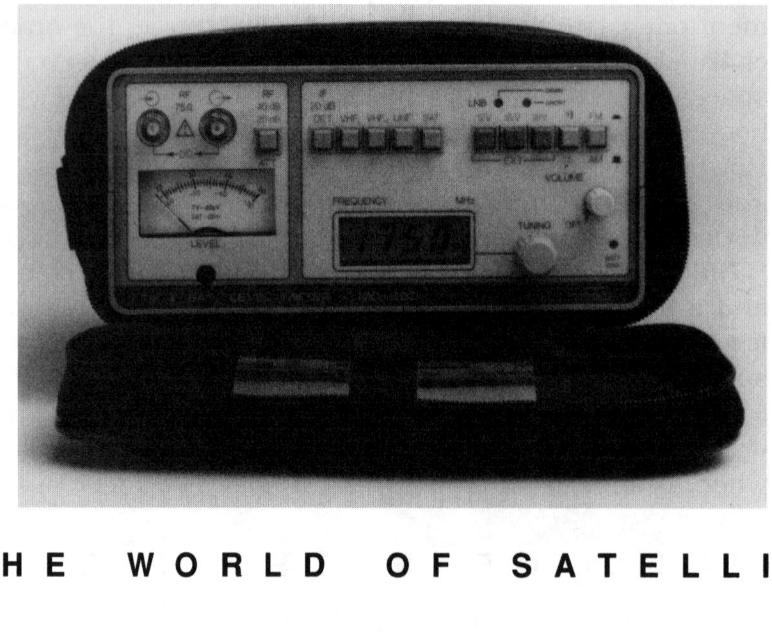

Fig. 9-9. A signal meter, which measures in dB, is required for setting the power output of the RF modulators and strip amplifiers. The model shown can be used for both broadcast TV and satellite signal measurements. (Courtesy Promax.)

*Fig. 9-10.
dBmV to μvolt
conversion
chart.*

dBmV	μVolts	dBmV	μVolt
-40	10.00	8	2,512
-39	11.22	9	2,818
-38	12.59	10	3,162
-37	14.13	11	3,548
-36	15.85	12	3,981
-35	17.78	13	4,467
-34	19.95	14	5,012
-33	22.39	15	5,623
-32	25.12	16	6,310
-31	28.18	17	7,079
-30	31.62	18	7,943
-29	35.48	19	8,913
-28	39.81	20	10,000
-27	44.67	21	11,220
-26	50.12	22	12,590
-25	56.23	23	14,130
-24	63.10	24	15,850
-23	70.79	25	17,780
-22	79.43	26	19,950
-21	89.13	27	22,390
-20	100.0	28	25,120
-19	112.2	29	28,180
-18	125.9	30	31,620
-17	141.3	31	35,480
-16	158.5	32	39,810
-15	177.8	33	44,670
-14	199.5	34	50,120
-13	223.9	35	56,230
-12	251.2	36	63,100
-11	281.8	37	70,790
-10	316.2	38	79,430
-9	354.8	39	89,130
-8	398.1	40	100,000
-7	446.7	41	112,200
-6	501.2	42	125,900
-5	562.3	43	141,300
-4	631.0	44	158,500
-3	707.9	45	177,800
-2	794.3	46	199,500
-1	891.3	47	223,900
0	1,000	48	251,200
1	1,112	49	281,800
2	1,259	50	316,200
3	1,413	51	354,800
4	1,585	52	398,100
5	1,778	53	446,700
6	1,995	54	501,200
7	2,239	55	562,300

CROSS-CHANNEL INTERFERENCE

With all power output levels equal, no one channel should overpower or "bleed" onto any other channel. On many occasions, however, there still may be some "cross-channel interference"—the superimposition of ghostlike images from one channel onto the video of another channel in the system.

Determine which channel is causing the interference by trying to recognize a logo or some other identifying factor. First try to eliminate the interference by turning down the output of the offending modulator or strip amplifier. Sometimes just a small tweak with an adjusting screwdriver will take care of everything, removing the interference. If that doesn't solve the problem, it is possible the strip amplifier itself is being fed an impure signal, which it simply amplifies and passes on through the system.

Since you are dealing with single channel amplifiers, the only thing you want going into them is the signal of the desired channel. Band pass filters are designed to let only one signal through, while blocking out all others. Made for specific channels, band pass filters receive the signal from the antenna, before it gets to the strip amplifier. The filter traps out any signal except the specified channel.

Although not designed for this task, you can sometimes

achieve effective results by using a notch filter, which is the opposite of a band-pass filter. The notch filter traps out specific channels or frequencies related to the interfering channel. If the channel 6 modulator is putting out an interfering signal that shows up on channel 12, place a channel 12 notch on its output. The channel 6 signal will pass through unaffected, but the channel 12 frequencies will be eliminated.

Cross-channel interference also can be the result of poorly shielded cables between the various pieces of equipment within the head end. The really careful system designer uses quad-shielded cable to eliminate this as a potential problem source.

You may find that the problem is not caused by cross-channel interference at all. An off-air TV channel, for example, with wavy lines running through the picture may be delivering an excessive signal level to the input of its corresponding amplifier. If the site is located many miles from a transmitter site, the signal received can be very weak. Or if the head end is relatively near the transmitter site, the signal level may be quite strong, 25 dB or more. Strip amplifiers are designed to receive signal levels within a range of about 10 to 15 dB. If the incoming signal is stronger than this, the amplifier will be overdriven and signal distortion will occur. Signal levels that exceed the recommended input range of the amplifier must be reduced to a permissible level. To reduce the signal level, you must use a balancing accessory known as an attenuator. Attenuators are rated by the number of dB that they will reduce the signal—a range like 3, 6, 10, 12 and 20. Attenuators can be stacked together in different combinations to get just the right reduction in signal.

CARRIER POWER

To decide what output level is needed for your application, one must first examine the design of the entire system. Determine how many TVs are to be supplied with a signal and the distance that signal must travel. A motel with short runs may have just a few TVs, or it may have two hundred or more. A small private cable system in a subdivision or resort complex my have only 100 televisions, but they may be spread out over thousands of feet of cable.

This brings into play the second factor: the distribution system. Modulators in the head end of a small motel may be able to supply enough carrier power for all the TVs without additional booster amplifiers to keep the power at acceptable levels. Once your distances have been determined, it may turn out to be more economical to use 35 dB modulators with additional amplifiers than 55 dB modulators on every channel. Understanding the distances involved and the total system requirements can help determine which modulator is best for the particular application at hand.

MIDBAND & SUPERBAND CHANNELS

Whenever an SMATV system will carry more TV channels than can be accommodated on VHF channels 2 to 13, the system designer must select additional channel frequencies to accommodate the extra channels to be provided. There are nine "midband" channels placed in a band of frequencies between VHF lowband channel 6 and VHF highband channel 7 and more than thirty "superband" channels between VHF channel 13 and UHF channel 14.

Most TV sets manufactured for the North and South American markets these days are "cable ready"—that is, capable of tuning into the midband and superband channels as well as the standard VHF and UHF channels. Those subscribers who do not have a cable-ready TV or a TV with the ability to tune to midband or superband frequencies must be supplied with a cable converter box. The set-top tuner receives the midband and superband channels and converts their signals to a signal that can be displayed on one of the TV set's regular VHF or UHF channels. In this case, all channel selection is done at the converter, or via the converter's hand-held control. Many videocassette recorders now offer tuners capable of receiving midband and superband cable channels. If desired, this type of VCR can be used instead of a cable converter box.

ADDING IN-HOUSE CHANNELS

SMATV systems are not limited to providing channels from local broadcasters or satellite programmers. Adding your own in-house channel can be as simple as connecting the output of a VCR to one of the modulators at the head end. The signal is then distributed throughout the private cable system just like any off-air or satellite signal would be delivered. Character generators, often used as message systems, also can be connected to one of the head end's modulators to provide text information to subscribers over one of the cable TV channels. A separate source, such as a local radio station, can use the TV channel's sound circuit while the text information is being displayed on each TV screen in the system. A TV camera on the front door to a building or at a nearby playground also can provide a video signal that could be distributed throughout an individual complex.

THE CABLE DISTRIBUTION SYSTEM

Once the broadband multiplex carrying all of the desired TV channels leaves the head end, the signal must be divided, or "split," several times before it reaches every TV set in the cable distribution system. Each time that the signal is divided, the signal level is reduced at that point in the system. Cable amplifiers must be added at various points along the cable line to boost the signal level back to an acceptable level.

An apartment complex, for example, may have an amplifier for every building or every few buildings, depending on the length of the cable run and how many TV outlets it has to feed. Cable systems that handle whole cities may have hundreds of amplifiers. On small private cable systems, these amplifiers are called "line extenders."

An engineer is usually responsible for the layout of large distribution systems. It is the task of the engineer to prepare a set of plans called a "strand map." The strand map depicts the interconnection of the cable lines, amplifier locations, and the TV signal outlets. The engineer looks at the entire system, calculates the cable loss and the number of TVs involved.

HARDLINE VERSUS RG6 CABLE

In smaller systems, such as at a hotel or motel, RG6 coaxial cable with a center conductor and an outer aluminum foil shield often is used as the main distribution or "trunk" cable. The excessive loss of RG6, however, becomes noticeable over long distances, and amplifiers will be required at periodic intervals to boost the signal level back to an acceptable level.

When designing for a small subdivision or any system requiring cables of great length, the coaxial cable used is much larger, generally about half an inch in diameter. Larger installations use a lower loss "hard-line" cable with a solid aluminum outer shield. Hard-line cable is stiff and somewhat difficult to work with. It must be handled carefully to avoid any "kinks," which will cause anything from an immediate short to signal reflections that create ghosts on the TV sets connected to the system.

Hard-line cable requires its own special connectors and tools for installation. Although it is flexible enough to make gentle curves in a ditch, sharp 90 degree bends must be done using right angle connectors. There also are special splice fittings, male input, and housing-to-housing connectors, or even connectors for terminating hard-line cable to a standard female "F" connector. There are several brands of hard-line connectors to choose from, with each requiring a different preparation technique during the installation process.

Systems feeding entire cities generally have cable runs suspended in the air from electric or telephone poles. Most private cable systems install their cable underground, a more aesthetic approach that does not require the leasing of space on local utility poles.

Underground cable is coated in a black plastic jacket. Between the jacket and the aluminum is a sticky substance that can best be described as a gooey mess! Although it does prevent moisture from entering the cable and causing problems, some kind of solvent will be needed to remove it from the installer's hands.

When it is time to tap off some signal, add an amplifier, or add a splitter, the cable comes back out of the ground and into a green metal enclosure called a "pedestal." These come in large and small sizes, depending on the amount of equipment they must contain. A big pedestal may hold an amplifier, a splitter, and/or combinations thereof. A latch on the front is provided for a padlock. Pedestals are vented to dissi-

pate heat and prevent condensation from affecting the electronics.

SELECTING A DISTRIBUTION STYLE

Whether you are designing your own distribution system or working from a plan provided by an engineer, it pays to be familiar with the different methods used to route a broadband signal to multiple TV sets.

The Home Run Method. Probably the simplest method is called the "home run." Here the cable for every single TV comes all the way back to a single distribution point, usually at the head end. You usually will find that the home run method is used for serving small systems with just a few TV sets. The advantage of the home run method is that fewer splice points are required, eliminating potential points along the cable where a bad connection and loss of signal might occur. The home run method also allows the system operator to easily control which signals are sent to each subscriber—an essential ingredient in any simple pay-per-view delivery system.

The Series Distribution Method. Another method often encountered in older motels is the "series distribution" system. In this case, a single cable leaves the head end, and, at every TV along the way, a tiny bit of signal is siphoned or "tapped" off, with the cable continuing on its journey to the next TV set in the system. The cable may zigzag up and down from the bottom floor room to the top, and then head back down again. Series distribution systems involve a lot of connections, and therefore there are many potential points for problems to occur. This method has almost been entirely phased out of use.

The Trunk and Feeder Method. The most common technique in use today by private cable systems is the "trunk

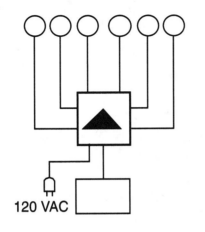

HOME RUN CABLE SYSTEM METHOD

120 VAC

Fig. 9-11. The home run method for the cable distribution system.

and feeder" method, also sometimes referred to as a "tree and branch" system. A main trunk line, usually half-inch diameter hard-line or larger, carries the multiplexed signal from the head end all the way to the end of the line. Along the way, feeder lines are split off from the main trunk, with each feeder line carrying enough signal to drive the TV sets on that branch. Each

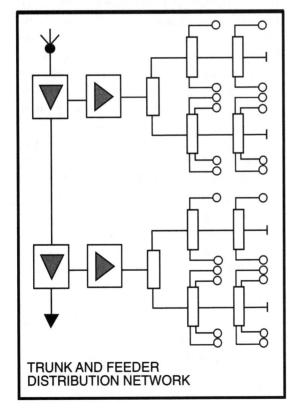

TRUNK AND FEEDER
DISTRIBUTION NETWORK

Fig. 9-12. The "Trunk and Feeder" method for the cable distribution system.

feeder line actually can supply a great number of TVs, with amplifiers or line extenders added as needed to complete each feeder line off the main trunk.

AMPLIFIERS NEED POWER

Nearly every distribution system will need some amplifiers to periodically boost the signal down the line. In some cases, electricity will be available at each amplifier location, allowing the amplifiers to be plugged into the nearest AC outlet. This is done often in hotel/motel systems, or even in systems serving apartment buildings. The use of AC powered amplifiers, however, is only practical when there is one central electrical system that is controlled by the building or system owners.

More often, the amplifiers must be powered remotely, through an AC voltage that is carried piggyback on the distribution cable right along with all the TV signals. A power supply at the head end produces the 30 to 60 volts AC required to power all of the amplifiers in the system.

A "power inserter" is required to place AC power onto the cable. The power inserter has a small rectangular aluminum housing, with ports marked "RF IN," "AC IN," and "RF/AC OUT". Each input and output inside the power inserter is fused as a safety precaution and to give greater control over the flow of electricity. The fuse on the RF input should be removed prior to applying power so that no AC can find its way back into the rack of equipment.

TAPPING OFF THE SIGNAL

The signal level on the main trunk's hardline may be around 45 dB. To deliver 10 dB of signal to one of the TV sets in the system, 10 dB of signal must be siphoned off the trunk line while sending the remaining amount of signal

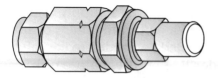

CABLE TERMINATOR

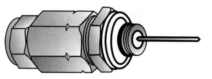

CABLE SPLICE

Fig. 9-13. Hardline cable connector types.

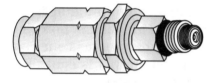

MALE F CONNECTOR

PIN TYPE CONNECTOR

FEMALE F CONNECTOR

FEED THROUGH CONNECTOR

to other subscribers down the line. A power splitting device, called a "tap", is like a water valve which is only partially open. It is used to divide the signal into two or more streams.

Taps are rated in dB. With 45 dB of signal on the line and 10 dB needed at the TV, you would install a 35 dB tap. At the female "F" fitting on the tap, there will be a signal output of 10 dB, with the remaining 35 dB going down the line to the next tap location. Although a few dB may be lost as the signal passes through the tap and the coaxial cable between the tap and the TV set, there will still be sufficient signal to deliver a good, sharp picture. Keep in mind when determining the tap value, however, that the signal may be split at each subscriber location to feed a second TV in another room. The tap also does the important job of removing the AC voltage from the signal that might otherwise cause damage to the subscriber's TV set.

INSTALLING HARDLINE CONNECTORS

The installation of hardline connectors requires special tools and techniques. The installation basics, however, are just a matter of hot and ground—a center conductor and an outer shield separated by a foam dielectric. In all cases, the center conductor, or "stinger," extends past the aluminum shield by two or three inches.

The cable is cut to length using either a hacksaw or a pair of cutters. On underground or flooded cable, the next step is to peel back the outer plastic shield. A tubing cutter is used to lightly score the aluminum outer shield all the way around. The location of the scoring determines the length of the stinger, so make your mark accordingly.

Grasp the end of the aluminum with a channel-lock or pair of pliers. A side-to-side bend will snap the shield loose. Pull straight up and slide it right off the foam dielectric. If it proves to be difficult to pull the shield off, you may have scored too deeply with the tubing cutter. When installing underground cable, take a rag and some solvent, such as paint thinner, and wipe down the outside of the aluminum to remove any residue so that your connector makes a good ground connection.

Your next move depends on the style of connector used. Some connectors are ready to go right on, but these are considered "B Grade" because they do not offer the degree of shielding that helps to insure no signal leakage.

"Grade A" connectors have a sleeve that slips down inside the aluminum shield, effectively enclosing the center conductor. To accomplish this, remove the foam dielectric approximately an inch from inside the shield. This is done with a "coring" tool. The center conductor goes up through the middle while the tool is twisted like a corkscrew. The foam comes out and the connector is ready to go on.

The most common type of hardline connector in use is called an "Entry" connector. It joins the hardline cable to just about any and all pieces of hardware you will ever use. This connector comes in two pieces. The bottom piece first slides over the cable. A wrench is used to thread the top half into the body of your hardware. The two halves are then threaded together. As they are tightened, a snug fit is made onto the cable. Grooves in the connector bite into the cable's outer sheathing, ensuring a good ground connection. Don't overtighten; you could kink or damage the hardline cable. A big piece of heatshrink tubing over the whole splice helps to seal out moisture.

Use an ohmmeter (there should be no voltage on the line) to check continuity and for shorts. If you make the stinger too long, it can come in contact with the body of the connector or tap, causing a direct short. It is always a good idea to check the cable for shorts before you apply power to the line.

THE DROP
The run from the tap to the house is known as a "drop." Standard RG6 or RG59 coaxial cable is used for this purpose. If the drop runs through the air, the coax will have a "messenger"— a steel cable bonded to it that supports the cable's weight at each end. In underground systems, the coaxial cable comes out of the ground just outside of each house or building served by the system. Use a "flooded" cable, which has a sticky substance between the shield and the dielectric to keep out moisture.

At the house, you'll need a ground block connected to a ground rod pounded several feet into the ground. The ground block has a double female F connector with a screw terminal for a ground connection. This provides a measure of protection from lightning and power surges.

PROGRAM DELIVERY OPTIONS
Many systems will offer multiple packages or "tiers" of program channels: basic services, such as off-air channels, satellite Superstations, news and sports services and premium movie channels. Although all subscribers will want access to the basic services, not everyone will be willing to pay the additional subscription cost for a premium movie channel or a pay-per-view event.

Since the premium movie channels are multiplexed onto the cable along with everything else, the system operator must have a way to block or "trap" these services before they enter the homes of subscribers that are not paying for the additional services.

A trap is a specialized filter that notches

out the unwanted channel or channels. It is installed in the pedestal on the output of the tap, in series with the coaxial cable feeding the customer's TV set.

One technique is to use a single trap to feed the signals into multiple output splitters. All houses or apartments subscribing to a particular tier are connected to a common splitter. These traps require more signal output from the tap to compensate for the extra splitter loss.

Still another method is the reverse trap. Here a special device at the head end places an interfering carrier on the video signal. At the home, the audio is present, but the video appears to be scrambled. A trap is installed in the pedestal that strips off the carrier and returns the video to normal.

One of the most rapidly growing areas in private cable distribution is Pay Per View, the ability to sell programs on an event-by-event or movie-by-movie basis. Until recently, the cost of this technology was too expensive for any except the largest cable operators. New cable distribution systems are now available that use inexpensive personal computers equipped with specialized software, which control special cable converter boxes.

HARDLINE SPLITTERS

Selecting the right location for installing your hardline splitters is one of the keys to planning an effective cable distri-

bution system. The hardline splitters divide the signal, allowing you to route the signal down different roads or directions. They also serve as the insertion points for any amplifiers or "line extenders" that are required to amplify the signal back up to acceptable performance levels.

A line extender will increase the signal by about 25 dB. Whenever the signal level drops down to about 20 dB, it is time to add a line extender. That returns you to a healthy 45 dB, and you are ready to continue. The signal strength meter with a calibrated readout in decibels is an essential tool for knowing exactly how much signal power you need at any point in the system.

The system designer usually figures this all out on paper in advance of construction, factoring in variables such as the signal loss on the cable per

Frequency (in MHz)	RG-59U	RG-6U	RG-11	.412	.500
	Attenuation per 100 feet (in dB)				
5	0.81	0.61	0.36	0.19	0.16
30	1.45	1.17	0.75	0.48	0.39
50	1.78	1.44	0.93	0.62	0.50
108	2.48	2.02	1.30	0.92	0.75
216	3.49	2.85	1.83	1.31	1.07
240	3.68	3.00	1.94	1.39	1.14
270	3.91	3.19	2.06	1.48	1.21
300	4.13	3.37	2.17	1.56	1.28
325	4.31	3.51	2.27	1.63	1.34
350	4.48	3.65	2.36	1.69	1.39
375	4.65	3.79	2.44	1.76	1.44
400	4.81	3.92	2.53	1.82	1.49
450	5.13	4.17	2.69	1.94	1.59
500	5.43	4.42	2.85	2.05	1.68
550	5.72	4.65	3.01	2.16	1.77
600	6.00	4.87	3.16	2.26	1.86
650	6.27	5.09	3.30	2.36	1.94
700	6.53	5.29	3.44	2.45	2.02
750	6.78	5.50	3.58	2.55	2.10
800	7.03	5.69	3.71	2.64	2.18
900	7.50	6.07	3.97	2.81	2.32
1000	7.95	6.43	4.23	2.98	2.46

Fig. 9-14. Cable attenuation chart.

hundred feet, loss due to splitting the signal for each feeder line, and the number of TV sets involved. It is possible to determine the proper values as you go along, keeping track of the signal level and adding a line extender whenever necessary to maintain acceptable performance levels. But first, get your calculator out, arm yourself with a few statistics, and get a rough idea of what materials you will need. Figure on a little signal overkill when working out the estimate for larger installations.

Keep in mind that channels higher in frequency will show more loss than low band channels like VHF 2-6. In a system that uses midband frequencies, channel 13 will show the most loss. Move up into the superband frequency range, and the loss is even greater. For this reason, the dB levels are often "tilted" in the direction of the highest channels.

When the power output is set at each amplifier in the system, the signal level of a high-frequency channel may be 5 dB or more than the channel 2 signal level. With cable loss, by the time the broadband signal reaches the next amplifier, the tilt will have equalized. The output of each additional amplifier in the distribution system is tilted throughout the system to insure that channel signal levels are equal at all delivery points.

OTHER HARDLINE SPLITTER USES

One of the most important pieces of hardware in a private cable system is the two-way splitter. The splitter does the obvious job of dividing the main signal so that it can be sent in two different directions. The hardline splitter, however, has some other potential uses as well. The splitter also can serve as a convenient housing for other electrical circuits that are needed to make the distribution system perform correctly.

Hardline Splitter as a Power Inserter. The hardline splitter can be used at the head end as a "power inserter" to place AC voltage onto the cable distribution system. This simple but vital function is found in all systems utilizing AC line powered amplifiers.

Installation is fairly straightforward. Each port on the inserter is clearly marked. On one side is the input for the signal from the equipment rack, coming from the output of the combiner. You can count on it nearly always leaving the head end rack as a piece of RG-59 or RG-6 cable. The connection for the AC power is located at the other input. It also comes to the inserter as a cable from the power supply, either on a coaxial cable or a piece of hardline.

First, the appropriate style connectors are installed on the inserter, a hardline entry connector. With the two input signals connected, the output of the inserter is now a multiplexed signal containing all of the many channels from the combiner, plus the AC voltage.

The Hardline Splitter as a Splitter. Distribution systems consist of a main trunk line, which at times is divided or split, routing the signal down separate paths. If used as a pure splitter, the signal is divided into two equal halves. But more often, the trunk cable serves as the primary line with other shorter legs tapping off to feed smaller groups of TV sets. Therefore, only a small portion of the signal is removed from the main trunk, while the balance continues on down the line to the next distribution leg or amplifier. The tap off port will be labeled in reference to the amount of signal siphoned away, something like 8 or 12 dB. The tap also can be used to remove an unwanted amount of signal before reaching a distribution amplifier

CHAPTER 9 : MULTI-DWELLING SIGNAL DISTRIBUTION SYSTEMS

or line extender so that you do not overdrive the amplifier and create distortion.

Hardline Splitter to Control Voltage. In many cases, the shorter distribution legs will not have any amplifiers. The signal runs down the feeder line until all TV sets on that leg have been served. Since this feeder line has no need for AC power, a splitter equipped with a removable internal fuse can be used to remove AC power from that portion of the line while allowing the RF signal to pass through.

This can be helpful for a number of reasons. Power consumption is reduced by not sending electricity down a leg of cable that has no need for it. Although the amount of electrical consumption may be small, if several lines are involved, this can add up. Also remember that this unnecessary AC consumption will take place 24 hours a day, every day throughout each year.

Another reason for using a splitter with a removable internal fuse is for added convenience during servicing. If there is a bad amplifier which needs to be replaced, or even more serious, a damaged cable which must be spliced, the repair task will be much easier if the AC is removed from the affected feeder line before the work begins. If the AC power is turned off at the head end, voltage will be removed from the entire system, disrupting everyone's service. It is much better to remove power only from the feeder leg which needs servicing, leaving all other legs operational.

As might be expected, the internal fuses also serve to protect hardware and equipment. Should a power surge enter the line due to an accidental short or a nearby lightning strike, the fuse will blow, stopping the power surge dead in its tracks.

DISTRIBUTION LINKS

In today's urban environment, developers must often find a way to link properties or buildings separated by great distances or apparently insurmountable obstacles. A popular although expensive solution can be a microwave or even laser link. The transmitting and receiving antennas must be kept within the line of sight and accurately aligned. When a direct cable connection is impossible, either due to a natural or manufactured obstruction, or a legal "right of way," the wireless route can be the perfect solution.

INSTALLATION

For the cable operator running overhead lines, the installation process is fairly straightforward. The trunk cable comes in from one end and the two (or even three) outputs exit from the other.

However, private cable installations almost always have underground distribution systems. All cables, the input as well as the outputs, come from the same direction, straight out of the ground. The input and outputs on the splitter are on opposite ends. One must make a 180-degree turn to reach the ports on top of the splitter, all within the confines of the pedestal. Even the slightest bend in the hardline cable, if not performed carefully, can cause a kink. At worst, the aluminum outer shield collapses, shorting out the center conductor. Or a kink can create a "ripple" in the signal that can create ghosting on TV sets connected to the line.

To accomplish a 180-degree turn in a space of only a few inches, special "right angle" connectors must be used. They may vary slightly in the length of their center conductor. Two of them must be used together (90 + 90 = 180) to complete the turnaround.

Let's look at an example. Assuming

146   **THE WORLD OF SATELLITE TV**

the cable lengths exiting the ground permit this, the most logical method of installation is to have both outputs enter the splitter directly from the bottom, a straight shot. Only the input cable must go through the turn. The cable comes up, enters a right-angle connector, and a second right-angle connector completes the curve.

The right-angle connector is female on one end and male on the other. The cable is stripped back leaving a "stinger" that is a couple of inches long. An entry connector is placed onto the cable before attaching it to the female side of the right angle. The output end of the connector has its own stinger, which is fed into the next right-angle connector, splitter, amplifier, or whatever other component is next in line.

A secure and positive connection is made with the stinger by means of a setscrew. The screw is accessed through the removal of a small cap, and is tightened down with a common flat-head screwdriver.

Installation of the stinger in the right-angle connector requires special attention. Because the connector makes such an abrupt turn, you must cut the length of the stinger to the proper length so that it does not come in contact with the outer case or "ground" of the second connector. Once the connector is installed, check it with an ohmmeter, placing one probe on the setscrew, and the other against the outer case. If it shows a short, take the connector apart, shorten the center conductor, and reassemble the unit. It should be just long enough to slip past the setscrew by about a quarter of an inch. Cut it too short and you risk the chance of making a poor connection as fluctuating temperatures can cause electrical contacts to separate whenever the connection is marginal.

In some cases it may not be necessary to make a full 180-degree turn. Each splitter port has two points of entry, 90 degrees apart from each other. Through one opening is the entry from the stinger. The 90-degree counterpart is where the setscrew resides. These can be alternated, as the screw can be pivoted, placing the setscrew at the stinger entry position and vice versa. The stinger entry is now on the side so only one right angle connector is needed to complete the turn. The ability to alternate or choose between a side and end point of entry can be found on most hardline equipment, including amplifiers and taps. Basically, it increases your installation options, and can be helpful when installing many pieces of equipment within the limited confines of a pedestal.

Familiarity with the hardline splitter is essential for the installation of any major distribution system. Having a better understanding of the various types and uses will help you during the design stage and when ordering parts.

DIGITAL SMATV SYSTEMS

Many SMATV systems already receive digital satellite TV services and distribute them to their subscribers. At the present time, each digital satellite signal is typically decoded at the head end, converted to standard video and audio signals, and then passed through an RF modulator that puts the video and audio onto a standard analog TV channel. Each digital TV service, whether satellite or off-air, will require its own dedicated receiver, decoder and associated RF modulator. When digital TV signals are distributed in this fashion, the various benefits of high-quality digital TV and sound never reach the SMATV subscriber due to the fact that the SMATV head end has converted the digital signal into an analog signal prior to distributing it to subscribers.

Chapter 6 briefly discussed the Digital Video Broadcasting (DVB) Group and its efforts to develop a worldwide standard for the broadcasting of MPEG-2 digitally compressed video and audio signals. One of the key reasons for establishing further agreements concerning use of the MPEG-2 digital compression standard was to provide for the cross-platform portability of digital video signals. Any DVB-compliant signal can be transported from one TV environment to any other without requiring any changes to the original digital bit stream. For example, a digital cable TV head end or a terrestrial DTV broadcast station can receive a DVB-compliant digital satellite signal and redistribute it without needing to decode the transmissions first.

A single 36-MHz-wide satellite transponder may carry a digital multiplex containing as many as six TV program services with associated sound channels as well as other auxiliary audio and data services. A digital SMATV system can receive this digital multiplex and pass it directly down the cable to its subscribers. To accomplish this, a digital channel processor—called a "transmodulator"—is installed at the SMATV head end to convert the wideband digital satellite signal into a narrower signal that can be passed over a single cable TV channel.

All DVB-compliant satellite signals use a form of modulation called QPSK. The head end's transmodulator converts the QPSK signal into an equivalent QAM (for Quadrature Amplitude Modulation) signal that can squeeze the information contained in the original satellite signal into a 6-MHz-wide cable TV channel. In this case, all SMATV subscribers will need a digital set-top box that has been

designed to process QAM modulated signals, which allows them to have high-quality digital TV and audio signals delivered directly to their homes.

The new DTV standard for digital TV broadcasting in the USA allows each terrestrial TV station to simultaneously transmit as many as five STV (for standard TV) or two HDTV (high definition TV) services over a single terrestrial TV channel frequency as well as auxiliary audio and data services. The digital cable head end also can seamlessly relay all of the signals contained in the terrestrial broadcaster's digital multiplex to each subscriber's home in a high-quality digital format that can be decoded by each subscriber's QAM-compatible set-top box.

The cross-platform portability afforded by the DVB standard means that the complexity of the SMATV system head end can be reduced dramatically. Instead of needing a separate receiver or IRD for each satellite TV service, the digital head end merely needs a separate transmodulator for each digitally modulated satellite transponder.

For example, a digital SMATV head end with just twelve transmodulators could potentially deliver more than ninety TV services to its subscribers using just twelve cable TV channels. An analog-based SMATV would require ninety receivers and ninety RF modulators to provide an equivalent service over ninety cable TV channels.

There are other ramifications that should be considered, such as the ability to use less expensive coaxial cables and other distribution components. At the other end of the continuum, the digital cable TV plants for major urban areas can now offer their subscribers access to hundreds of high-quality digital video, audio and data services.

The high cost of implementing a digital SMATV head end may make it an impractical solution for small system operators to implement. In this case, there is an alternative method for providing full digital service directly to SMATV subscribers. Some DBS and digital DTH satellite operators now offer multi-dwelling SMATV packages that allow a single head end to distribute their entire package of digital DTH services. Each SMATV subscriber is equipped with a digital satellite IRD, with all of the IRDs in the system connected to a single antenna at the head end that is pointed at the DBS satellite constellation of choice.

What's more, only a single coaxial cable is required to link each IRD to the antenna. The multi-dwelling distribution system uses a special dual-polarization LNF that produces a wideband IF output. The block IF frequency band, which ranges from 950 to 1950 MHz, contains signals using both satellite polarization formats. The lower range from 950 to 1450 MHz contains all signals of one polarization, and the upper range from 1450 to 1950 MHz contains all signals of the opposite polarization. Each IRD in a multi-dwelling system has been designed to receive and process the entire range of signals contained within the 950 to 1950 MHz IF output provided by the LNF.

The main advantage for the system operator is that many of the components normally found in a traditional SMATV head end, such as satellite receivers, descramblers, and associated RF modulators, are no longer required. The operator can continue to offer all of the off-air TV channels that are available locally. These terrestrial TV signals can be multiplexed onto the same coaxial distribution line using the VHF and/or midband TV frequency channels located below 950 MHz, the point at which the satellite signal distribution begins. The SMATV operator can also use a larger dish than what is normally required for a home installation in order to reduce the frequency and duration of any rain outages.

The downside to this digital distribution method is that the wideband 950 to 1950 MHz signal will require more expensive line amplifiers, splitters and other in-line devices than what traditional SMATV systems typically employ. Increased cable attenuation losses will also come into play, which will require the use of more expensive, low-loss coaxial trunk lines and perhaps a greater number of line amplifiers, depending on the distances to be covered. Each subscriber will also need to purchase one or more digital satellite IRDs, which raises the initial installation cost per customer for signing up. The end result, however, is that each subscriber will obtain direct access to high-quality digital TV and audio services.

MULTI-CHANNEL MULTI-POINT DISTRIBUTION SYSTEMS (MMDS)

Multi-channel Multi-point Distribution Systems (MMDS) are used at many locations worldwide to deliver pay-TV services to subscribing households. In some countries, MMDS is commonly referred to as "wireless cable" because of the technology's ability to deliver a significant number of television programming channels to any customer's home within a given local area, without the use of a cable. Instead, the MMDS channels are sent "over the air" and can be viewed by any subscribers equipped with a corresponding receiving antenna and frequency converter.

There are many similarities between SMATV and MMDS distribution systems. Both technologies receive their program material via satellite. The programming is then reconfigured at the satellite receiving "head end" and then relayed as a package to local subscribers. Both systems also can be designed to offer individually priced sub-packages of programming, called "tiers," to their subscribers, or even provide Pay-Per-View movies and special events that can be ordered on demand. Both delivery systems require that a technician come to each subscriber's residence to install the basic receiving equipment. The major difference lies in just how the services are delivered to the customer.

SMATV systems require a significant investment of time and money on the ground to build a wired grid that passes each and every residence or within any local coverage area. MMDS transmits a signal directly over the air, with good line of sight coverage for ten to thirty miles, depending on terrain. Since an MMDS service does not have to design, build and maintain a costly cable system, the overall cost of establishing a service in any given area is typically one-third to one-half the cost of a comparable cable distribution system. MMDS services can also be established more quickly than wired distribution systems, and can reach all of the potential subscribers within the intended coverage area right at start-up. If reliable, high-quality components are used at both the transmitting and receiving end, there is a substantially lower level of signal outages, which results in a higher level of customer satisfaction. Since MMDS is less complicated than most wired delivery systems, there are less components that can fail and require servicing.

MMDS, however, does have several disadvantages compared to SMATV. MMDS is not a cost-effective way to provide service to residents within a single building or complex. Moreover, there are licensing requirements in most countries that must be completed before operations can begin.

The available number of MMDS channels in any given area is limited, which means that fewer channels of programming can be offered to subscribers. More costly equipment is also required at the customer premise. MMDS signals also are limited to the line of sight. This may not be a problem in a low-rise environment. In a high-rise environment, however, the MMDS operator must work to fill nulls and voids caused by terrain and/or buildings in the service's over-the-air signal coverage.

Some systems may require the use of a "back feed," a repeater which sends the signal back toward the transmit tower to fill in an area blocked by hills, tall buildings or other obstructions. (To provide isolation between the two signals, the back feed transmits using the opposite sense of polarization.)

Some of the more advanced cable TV distribution systems in North America and elsewhere also now are providing their subscribers with interactive capabilities, such as home shopping. The cable acts as a two way street where information can flow in either direction. Since MMDS is a wireless service, there is no way for data to flow from the subscriber back to the system operator short of using the telephone line.

MMDS FREQUENCY PLAN

The frequencies assigned for MMDS operations in the U.S. consist of thirteen 6 MHz-wide channels in the 2150 to 2162 MHz and 2500 to 2686 MHz bands. Three different channel groups are available: MDS 1 & 2; E1~E4 and F1~F4; and H1~H3. Other countries in the region have their own frequency assignments and channel groupings. Overseas readers should contact their national telecommunication authorities for further information.

Like the microwave satellite signals previously described, MMDS transmissions are signals that exhibit many of the characteristics of visible light. Microwaves travel along the line of sight and therefore can be blocked by hills and other rough terrain or by buildings or other manufactured obstacles. It is essential that both the transmit antenna and any associated receive antennas be positioned so that they have a clear view of each other.

THE TRANSMIT SITE

The MMDS operator must construct a delivery system that can deliver a strong enough signal to each subscriber's home so that a good picture, one that the customer will be glad to pay for, is received. The MMDS system operator will construct a head end system similar in many respects to the head end previously illustrated. The video and audio output of each satellite TV receiver, VCR, or other program source feeds into a Radio Frequency (RF) channel modulator. Each RF modulator output is then up-converted to microwave frequencies and greatly amplified by an MMDS transmitter (also called an amplifier by some manufacturers), which outputs a signal that typically ranges from 10 to 100 watts, depending on the local terrain and the size of the area to be served. The outputs of these transmitters first goes into a signal combiner, which connects to the transmit antenna via a transmission line. The transmitting antenna site must be located at a high, unobstructed location to overcome any obstructions to the signal, such as rough terrain or tall buildings. As the power requirements of the MMDS system increase, so does the associated cost of the transmitters, the combiner,

the transmission line, and the transmit antenna.

THE RECEIVE SIDE

From the transmit site, the MMDS signal radiates through the air to reach all locations within the primary coverage area of the system. At each subscriber's home, a relatively small and inexpensive receiving antenna picks up the signal. For adequate reception of the microwave signals, the receiving antenna must have a clear line-of-sight view of the transmit antenna. Unlike lower frequency radio waves, which can penetrate the walls of a building or nearby foliage, the signals used for MMDS transmission are limited by any obstructions between the transmitter and receiver. For this reason, the antenna typically is located on a tall mast anchored to the house.

A small, medium, or large antenna is selected for each subscriber's home, with the size selected depending on the distance from the transmit site. The longer the distance the greater the antenna size required to receive a good picture. A broad beam antenna typically is used at more centrally located homes, while a directional antenna is more often used at locations towards the edge of the desired coverage area.

After arriving at the antenna, the MMDS signal is shifted or "downconverted" to a lower band of frequencies (either VHF, UHF or midband) that are compatible with the TV set's channel tuner. To minimize signal loss, a weatherproof downconverter often is mounted directly behind the receiving antenna. The output of the downconverter is then routed into the home via coaxial cable. This cable may connect to the antenna input of a TV set or a set-top converter.

A set-top converter is required whenever the MMDS transmits encrypted program services, or whenever there aren't enough available VHF or UHF TV channels to handle all of the MMDS program services. If the downconverter converts some channels to midband frequencies, cable-ready TV sets can receive all of the unencrypted channels without the aid of a set-top converter.

SIGNAL SECURITY

There is a certain level of security built into any MMDS system because of the special equipment required to receive the microwave signals and convert them to channels which are viewable on any standard TV set. Security can be enhanced by encoding some, or even all, of the channels. The degree of encoding can range from simple sync suppression or video inversion to more sophisticated addressable systems that are controlled by computers. Addressable systems control what can be viewed by each subscriber's converter, thus providing a means by which the system operator can offer both basic and premium programming packages, PPV events, and the ability to disable the converters of customers who fail to pay their monthly subscription fees.

It should be noted, however, that addressable converters cost two to three times what standard, non-addressable converters cost, so operators must weigh the additional benefits of enhanced security and the ability to deliver multiple tiers or pay-per-view against the additional investment per subscriber required. Addressable systems also use an expensive computer and computer software, as well as a special encoder that must be installed at the head end for each MMDS channel to be encrypted prior to transmission.

LOW POWER MMDS SYSTEMS

Low power, one-watt, broadband MMDS amplifiers recently have become available which combine the output of up to eight channel modulators and then upconvert and amplify the whole range of frequencies together. There is a significant cost saving to this design over using eight separate amplifiers and a high frequency combiner. Because of the lower costs involved, low power MMDS distribution systems can be used to provide services to smaller communities which may not have a large enough subscriber base to warrant the installation of a higher power system.

This new design approach also avoids the signal loss normally associated with high frequency combining after the amplifiers. A one-watt per channel system can therefore be engineered to achieve nearly the same signal level as an eight-channel conventional system using ten-watt amplifiers. Additionally, higher power levels per channel in the broadband amplifier can be achieved when fewer channels are combined. The same one-watt, eight-channel system will deliver two-watts per channel when six channels are combined.

INTERFERENCE PROBLEMS

The most common sources of strong interference to MMDS signals are caused by airport and weather radar systems that use S-band frequencies that are adjacent to the frequency band used for MMDS signal delivery. MMDS interference problems are often solved through the use of a very sharp pre-selector filter that reduces the level of the interfering signal before it reaches the receiving system's pre-amplifier. Some MMDS downconverters come with a built-in pre-selector filter. These filters also are available separately and can be added to any downconverter.

TROUBLESHOOTING

This guide to troubleshooting satellite TV receiving systems will provide you with an overview of the common problems you might encounter, how to identify them, and how to repair them. Many of the troubleshooting techniques described below use the "divide and conquer" method to isolate the system component that is causing the problem.

The specific steps that we need to take to troubleshoot and repair any satellite receiving system will depend on the type of system in use. Small fixed-dish systems, such as those that receive digital TV signals, will often experience problems that are dramatically different from those that affect systems equipped with large steerable dishes. The approach taken here is to first present the problems that are unique to small dish systems, immediately followed by a description of the problems with cables and connectors that are common to all DTH receiving systems, big and small. Additional sections on the troubleshooting of large steerable dish systems and the use of signal measurement instruments appear at the end of the chapter.

SAFETY FIRST

Before we begin troubleshooting any satellite TV receiving system, however, we should take note of a few important safety concerns. When working on any kind of electrical equipment, think Safety First! Almost all satellite TV IRDs and receivers run on 110 or 220 volts AC (depending on the national electrical standard of the country in which you reside), enough to kill a person. The lower DC voltages that run on the cable that connects your receiver to the LNB and actuator, while not enough to cause physical harm, can cause irreversible damage to expensive electronic circuitry if improperly wired.

The following safety precautions should be observed at all times:

>> Never take the cover off any electronic device while its power cord is plugged into an AC wall receptacle. If you must work inside a set-top box, be sure you know the proper procedures. Without proper test equipment any thoughtless action may turn your entertainment center into an expensive junk sculpture as well as void the warranty.

>> Keep extraneous metal objects and wires well away from any system component when your are testing or otherwise working on it; this will help prevent an accidental short circuit.

>> Never turn on a set-top box unless a fuse of the correct amperage is in the fuse holder.

>> Do not touch any components inside any electronic device while the power is connected.

>> Water is a good conductor of electricity; do not work on a set-top box or antenna actuator if you are wet or standing in a puddle.

>> Never work on your system when weather conditions indicate that a lightning storm may occur in your vicinity.

DIGITAL DTH SYSTEM PROBLEMS

With a digital satellite TV signal you either have a perfect picture or no picture at all. It is like a light switch: either on or off. If one day your digital receiving system stops receiving a picture, you will need to use some form of signal tuning meter to determine the source of the problem. Many digital IRDs come with a built-in signal meter with a readout that is displayed as an on-screen graphic. It is a good idea to record the signal level at the time of installation so you can compare it to the level obtained if anything goes wrong with the system.

Rain Fades. There are some problems that no amount of troubleshooting can ever fully resolve. Reception problems caused by rain fades are inherent to the design of all Ku-band satellite delivery systems. Rain, fog or even rain-filled clouds rolling overhead will reduce the intensity of Ku-band satellite signals. The best we can do is to make sure that the system is operating at peak performance level. This will help to reduce the frequency and duration of any signal outage that occurs.

Loss of signal during a rain outage generally becomes evident in one of two ways. Some digital IRDs will display a freeze frame that represents the last video frame stored in the decoder's buffer circuit, while other digital IRDs will simply display a black screen with a "no signal" message superimposed on top of it. If your reception suddenly begins to cut in and out during light showers, this is usually a good indication that the system is no longer obtaining maximum signal levels. In this case, the obvious place to start is at the antenna.

Antenna pointing. If one day your digital DTH system stops receiving a picture, heavy winds may have pushed the antenna away from direct alignment with the satellite. Check the antenna to see if the marks made on the pole and the mount during the installation process still line up. If the picture fades in and out on windy days, tighten any loose bolts on the mount to eliminate any sloppiness or "play."

To check the antenna's alignment, have someone lightly push up or down, or left and right, on the rim of the antenna while you look at the IRD's signal level indicator. If your set-top box generates an audible tone to indicate signal strength, turn up the volume on the TV set, then go outside and lightly push up or down on the dish. If the audible tone and/or meter level goes up when someone pushes on the dish, the antenna is no longer pointing directly at the satellite.

The Low Noise Feed. The LNF is very susceptible to damage caused by the ingress of water as well as any lightning strikes that occur in your general vicinity. A lightning bolt doesn't have to physically hit the dish to wreak havoc on your system. Nearby lightning strikes can generate a transient voltage that can fry the sensitive microwave circuits inside the unit. An eight-foot ground rod pounded in next to the dish and an in-line ground block at the home's cable entry point provide some lightning protection for your system. Use copper ground straps to connect the dish and ground block to the ground rod. Check to be sure that these copper straps are providing your system with a good electrical connection. If any corrosion of these connections is discovered, disconnect the copper ground straps, scrape them clean with a knife, and then reconnect them to the system.

If you are able to receive some of the digital DTH services to which you subscribe but not others, this could indicate a problem with the LNF's polarization switching circuit or the 13/17 volts

DC switching voltage supply inside the IRD. To check this out, you will need to provide the DTH programmer with a list of the services that you cannot receive. The programmer can check to see if the missing services are all on satellite transponders of a given polarization.

Do not attempt to open the LNF because this will void its warranty. These units contain especially sensitive microwave circuits that are beyond the ability of most mortals to repair on their kitchen tables. If the LNF is receiving the proper voltage and there is no evidence of moisture in the connections, check to see if replacing the unit will correct the problem.

The Digital IRD. Check to ensure that the IRD's smart card is properly seated in the unit's conditional access slot and that your account is current with the program service to which you subscribe. Refer to the following section on problems with cables and connectors for additional troubleshooting information.

CABLES AND CONNECTORS

Before getting very far into troubleshooting cables and connectors, you will need to understand how to use a multi-meter or volt-ohmmeter, a basic electrical tool that can be used to measure voltages and to check for shorted connections or breaks in the wiring. The instruction manual that comes with any voltohm meter should quickly bring you up to speed.

To perform many of the operations recommended below, you will also need to purchase one or more double-barrel "female" F connectors from your local electronics supply shop. This inexpensive device can be used to temporarily replace electronic components that have been inserted into the cable, such as switches and line amplifiers.

Checking the RF modulator channel. Make sure that your TV set has been fine-tuned to the channel output of the RF modulator built into the IRD. Some TV sets may have to be programmed to receive the set-top box's RF modulator channel. You would be surprised how often this is overlooked. Many installations do not use the RF modulator connection. Instead, the video and audio outputs of the IRD connect to a VCR or directly to a TV monitor. The RF modulator only becomes essential if your TV set does not provide for direct video and audio input or if you are distributing the satellite signal to other TV sets around the home. If you are using the RF modulator and can receive a picture from the set-top box's video output but not from its RF modulator output, the problem is either in the RF modulator or in the cable connecting the IRD to the TV set.

Checking in-line amplifiers and switches. All set-top boxes deliver 13 to 18 volts DC up the coax to the LNB, even when the IRD is turned off. This helps prevent moisture from condensing inside the LNB housing and enhances LNB stability by minimizing any temperature variations. It also allows the IRD to receive data from programmers using encryption systems. This data keeps the IRD current with any technical changes that the programmer may make in regard to its transmission format.

Before you start, be sure to unplug the IRD from its AC receptacle so that there will be no power running up the line. This will protect the electronics from possible damage in the event that you accidentally short out the connections. If you accidentally blow the fuse by shorting out the cable, an accurate test cannot be made and damage to the outdoor electronics could also occur.

Start at the indoor unit and work your way out to the dish, checking any in-line amplifiers or other components along the way. Temporarily take any splitters or switches out of the system and use the double barrel F connector to connect the coaxial cable directly to the IRD. If the system works after this, one of the disconnected components must be the source of the problem.

Checking DC Power to the LNF. If the system has no in-line devices, or if you have completed the steps above without isolating a malfunctioning in-line device, you will need to check the IRD. Whenever plugged into an AC wall receptacle, the set-top box will generate 13 to 18 volts DC on the center pin of its IF input connector on the back of the unit. This DC voltage travels up the coaxial cable and powers the system's LNF or LNB.

If you have a DC/volt-ohmmeter handy you can check to see if this voltage is arriving at the LNF or LNB. Start by unplugging the IRD from its associated AC wall receptacle. Go out to the dish, unscrew the F connector from the IF output of the LNF or LNB. Make sure that no moisture has entered this F connector as it can cause a short in the cable.

Checking the cable connections. Look for any indication of moisture or its after effects. The interior of the connector should appear to be clean and its center conductor should not be touching the connector's outer metal sheath in any way. If you see any discoloration, such as the presence of brown or black goo that covers the inside of the connection, you will need to replace the old connector with a new one.

Any moisture here also can make the TV picture look jumbled, appear very weak, or even disappear entirely. Prevent this by applying a waterproofing compound around the connector or flooding the inside of the connector with a dielectric sealant.

If the outdoor connector appears to be fine, you will need to reapply power to the cable so that you can check to see whether or not 13 to 18 volts DC is arriving at the outdoor unit. Before you power up the IRD to check this out, be sure that the center conductor of the F connector will not come into contact with any metal surface on the dish, mount or elsewhere.

Plug the IRD back into the wall, set the volt-ohmmeter to measure between 13 and 18 volts DC. Apply the meter's red probe to the F connector's inner conductor and the black probe to the connector's outer metal sheath. If no DC voltage is present at the outdoor end of the cable, the problem may be a short or break at some point along the cable run. There is a special fuse or circuit breaker inside the IRD that protects the circuit that delivers a DC voltage to the outdoor unit. Any short in the line will blow this fuse, which is separate from

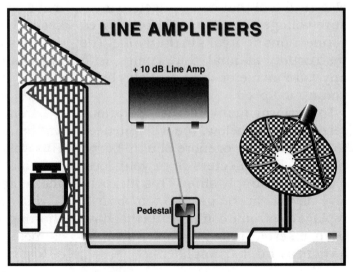

Fig. 10-1. Line amplifiers can be used to boost the signal level over long cable runs. Keep in mind, however, that they also are potential trouble spots for failure due to the ingress of moisture.

Fig. 10-2. LNB/Feedhorn potential trouble spots.

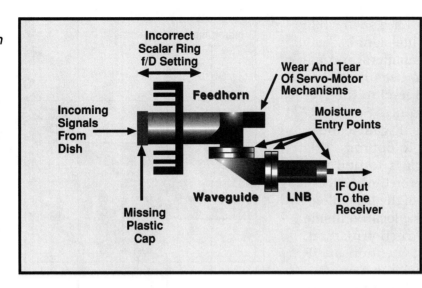

the unit's main fuse or circuit breaker.

Checking for short circuits and blown fuses. Go back inside the house, disconnect the IRD from its power source and unscrew the cable connector from the IF input port on the back of the IRD. Switch the volt-ohmmeter to its lowest range for measuring resistance. Apply the meter's two probes to the center conductor and outer metal sheath of the F connector. Low resistance between these two points indicates a short in the line. In this case, you will need to locate the source of the short.

Most often a visual inspection of the cable run will provide you with an indication of the source of the problem, such as a visible kink or break in the cable's outer plastic sheathing. If you discover any damage to the cable, you can cut out the damaged section, attach F connectors to each segment of cable and use the double barrel "female" F connector to rejoin the two segments. This cable splice point also must be thoroughly covered with a waterproofing compound to prevent the future ingress of moisture.

At this point, the IRD should still be unplugged from the AC wall receptacle. Locate and replace the blown fuse. Be-

fore you reconnect the cable, turn the receiver back on and see if the fuse blows once again. If it does, the short may have also caused a problem in the unit's power supply and the IRD will have to be sent back to the manufacturer for repair.

Checking for cable breaks. Lack of a DC voltage at the outdoor unit may also be caused by a break in the cable. In this case you will have to check the cable for continuity, not an easy thing to do if the cable is buried in the ground or permanently attached to the building. The easiest way to determine if there is a cable break is to purchase a sufficient length of RG-59 coax with F connectors on each end and use it to temporarily replace the length of cable that was installed along the dish. If the system works with the new cable in place, then you will need to replace the old cable run.

Checking the IRD signal input level. Every IRD manufacturer specifies a range of IF input signal levels for its product. A long length of cable between the indoor unit and the antenna may attenuate the signal arriving inside the home to a level that is insufficient to drive the receiver or IRD. Alternatively, the cable run may be so short that the resulting high signal level will overdrive the set-top box. This can inhibit the IRD's built-in decoder from effectively processing the encrypted satellite channels. High signal levels will also cause the analog IRD to generate noise that will appear in the picture. This noise looks just like the sparklies seen while receiving a weak satellite signal.

There are in-line amplifiers and attenuators that you can use to increase or decrease the signal level to the recommended range of values. Some set-top boxes provide an IF gain control that can be adjusted to optimize IRD performance. The trick is to find this control. Sometimes it is on the back, the bottom, or on the side of the IRD. Alternatively, it may even be located inside the unit. Check the instruction manual. If the manual doesn't mention an IF gain control you can ask the manufacturer if one is provided. The gain control should be adjusted so that the strongest channels almost peak the signal strength indicator on the IRD.

If every other aspect of your system checks out and you still don't receive a good picture or any picture at all, then it is time to call a local service center for assistance. Rural residents without local support may need to send their IRD back to the factory for repair.

LARGE DISH SYSTEM PROBLEMS

Antenna tracking. For a large, steerable dish that is mounted on a pole, move or "actuate" the dish until it reaches the satellite closest to the antenna's highest look angle. This is the satellite closest to the direction of true south for locations north of the equator or true north for sites south of the equator. If you push up on the antenna and the signal level goes up, the antenna's mount should be rotated slightly counterclockwise. If you push down on the antenna and the signal level goes up, the antenna's mount should be rotated slightly clockwise.

Be sure to make your adjustments in tiny increments and recheck the alignment by using the above mentioned procedure once again. If you continue

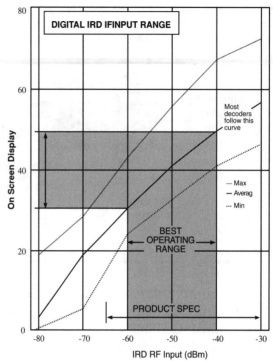

Fig. 10-4. The proper feeding and care for a digital IRD includes making sure that the signal level arriving at the IRD's IF input port falls within the specified ranges of optimum operating values.

to have an alignment problem, refer back to the alignment section of the installation chapter for further assistance.

Antenna actuators. The major culprit when it comes to problems with actuators is water! No matter what you do, water is going to get into the motor housing. Water accumulation can short out the connections to the actuator and blow the circuit breaker in the IRD or satellite receiver. This can easily be corrected by drying out the inside of the actuator and cleaning the connectors.

It is important to have a way for the water to escape. Make sure to follow the directions while installing the actuator, which usually place the incoming wires on the topside of the motor and the drain holes on the bottom. If there aren't any drain holes, remove the actuator cover and drill a drain hole in it.

If you don't move your dish frequently from one satellite to another, the arm can corrode and become stuck. The arm should be cleaned and lubricated periodically. Fully extend the antenna actuator arm, clean it with a solvent, and then lubricate the arm with waterproof "marine grade" grease that can withstand continual exposure to the weather.

If the dish won't move at all, a bad connection to the actuator's motor could be the cause of the problem. After making sure the parental lockout has not been activated, check to see if 36 volts DC is present at the back of the satellite receiver whenever the antenna is commanded to move.

Some IRDs provide 24 volts DC to the actuator motor instead of 36 volts DC. Providing only 24 volts will not affect the actuator motor's performance. Check the specifications provided in the owner's manual.

Next, check to see if this voltage is being received at the motor terminals inside the actuator housing. If it is not reaching the motor terminals, you may have to replace the actuator cable. If the voltage is present at the motor terminals and the actuator still doesn't engage, there may be an open winding in the motor which will need to be replaced.

All actuators have a circuit that generates pulses and delivers them to the receiver for counting. If the dish momentarily moves and then shuts down, the receiver didn't get a pulse from the actuator's sensor. An on-screen graphic or receiver front panel display will usually supply an "actuator error" message.

The type of sensor used in the actuators being manufactured today may vary. Some of these require 5 volts DC; others do not. Although a bad sensor is a common problem, it can easily be replaced. If your actuator is the type that uses a magnetic wheel, make sure that the sensor is close to, but not actually touching, the wheel. These magnets sometimes wear out and may need to be

replaced. If you determine that these components are good and the receiver still does not get a count, check the sensor wiring for continuity.

If your particular sensor requires 5 volts, make sure that the current is reaching the motor. If your receiver stops supplying 5 volts to the actuator, it would be a good idea to switch to a reed or other type of sensor which doesn't require voltage.

If the dish doesn't precisely locate a specific satellite every time, check the antenna mount for signs of any slop or play. Move the dish up or down by hand to discover which bolts may need to be tightened or what other antenna parts need to be replaced. If you need to realign the programmed locations for all satellites, it may not be necessary to re-enter the location for each individual bird. Many satellite receivers provide an automatic antenna realignment feature that can be used after reprogramming the location for a single satellite. Be aware, however, that this will realign your east and west limits as well.

If the IRD loses its memory, the unit's backup battery may not supply power to the memory circuits during power outages. The unit's microprocessor may also be blown. Refer to the owner's manual for further information about repair or replacement.

LNB and feedhorn. If your system uses a separate feedhorn and LNB assembly rather than an LNF, check the neoprene gasket that seals the joint between the feedhorn and LNB. The LNB should be tightly bolted to the feedhorn. If the mounting bolts are loose or if the gasket was improperly seated during installation, moisture can seep around this gasket and enter the feedhorn waveguide, the metal joint that links the feedhorn to the LNB. If moisture has indeed entered the waveguide, use a hair dryer to remove any condensation, then carefully reattach the LNB to the feedhorn flange so that the gasket completely seals the joint. Be sure to use ALL of the bolts that connect the two units together. On occasion, a feed will go bad and must be replaced when it will no longer pass signal. You should also look inside the feed opening to check for wasp nests. Manufacturers provide a plastic feed cap that should fit snugly over the feed opening to prevent the intrusion of wasps and other critters looking for a new home.

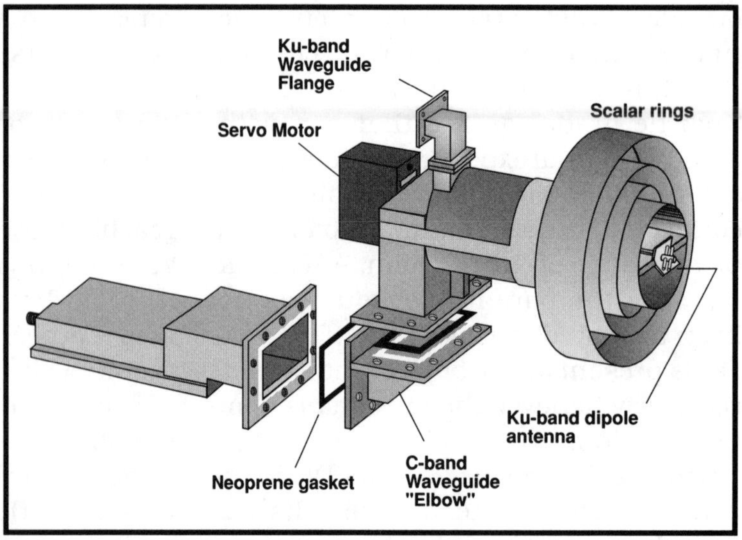

For those large dish systems that use an analog IRD, one indication of a bad amplifier stage in an LNF or LNB is the reception of clear pictures on the lower satellite channels but noisy pictures on the high end, or vice versa. In this case, you will need to replace the malfunctioning unit.

If you receive signals of only one polarization or signals of two polarizations at the same time and cannot adjust the skew to make the picture come in, there may be a problem with the servomotor on the feedhorn. The servomotor also can freeze up in winter. Moreover, the servomotor can wear out and must therefore be replaced.

The best way to check the servomotor is to bring the feed inside and hook it up directly to the back of the receiver. As you change the receiver's polarization or skew settings, the probe in the mouth of the feed should rotate in a clockwise or counterclockwise direction.

If the probe doesn't move, use a volt-ohmmeter to determine if the receiver is generating 5 volts DC whenever you

Fig. 10-5. Components of a dual-band feedhorn/LNB assembly. Note that the two mating flanges are potential trouble spots.

adjust the polarity. If not, the problem lies in the receiver's internal power supply. A few receivers have an internal fuse that protects this part of the power supply. Consult with the receiver manufacturer. If their product doesn't have a separate fuse, be sure to ask for a return authorization number and send the receiver in for repair. If you can't stand the pain of being apart from your receiver, buy an external polarization power supply.

If 5 volts DC is present at the feed and the probe still doesn't move, you will need to replace the servomotor. A replacement servo is relatively inexpensive. If the probe does move when hooked directly to the receiver, check the wires going out to the dish for a break or short in the line.

Remember, when installing a feedhorn on the antenna, it must be aligned so that the probe can swing 90 degrees from horizontal to vertical (or left-hand circular to right-hand circular) polarization without reaching its limit of travel. If you can't skew the probe beyond a good picture on odd and even channels on all satellites, the feed must be physically rotated until it is possible to do so.

Electronic ferrite feeds have no moving parts, so you won't be able to physically observe any movement during operation. Instead, use the volt-ohmmeter to determine whether or not the feed is receiving 18 volts DC. You can also check the windings to see if there is continuity across them.

The Analog IRD. If you have been on a long vacation or for some other reason your decoder loses its authorization, you can tune to any channel to which you subscribe and call the subscription service for an "instant hit." This should bring it back to life. If your IRD requires the use of a smart card, check to be sure that the smart card is inserted properly in the conditional access slot provided.

If the unit seems to have lost its memory and will not display its ID number, the battery in the decoder may have worn out. The battery is there to hold the memory intact whenever the power goes out or the unit is unplugged. Designed to last for several years, the batteries eventually wear out and will need to be replaced. The module will have to be sent back to the manufacturer or its authorized service center to have a new battery installed and your authorization keys reinserted. Opening the module will void the warranty.

Terrestrial interference problems. All C-band DTH systems also are susceptible to interference from terrestrial sources. In-band interference comes from terrestrial sources that operate within the same frequency range as the satellite or satellites that you wish to receive. Out-of-band interference can come from very strong signals that have a point of origin that is relatively close to the site location. Out-of-band interference is caused by one or more harmonics: a sub-multiple or multiple frequency to the one that you are actually receiving with your satellite dish.

Even if your initial site survey revealed no interference, there is no guarantee that the telephone company or some other user will not start transmitting through your area sometime in the future. On occasion, radiation from a local airport or TV station transmitter or even an improperly shielded cable TV junction box near the site location can wreak havoc on satellite TV reception.

With analog TV receiving systems, microwave interference typically appears as noise pervading a few of the satellite channels. In extreme cases, when the source of the terrestrial interference (TI) is strong and nearby, it can even wipe out a channel or a whole block of channels on one or more satellites. Some channels may be blacked out or have sparklies that simply won't go away, while other channels will look just great. A dead giveaway that TI is present is indicated when certain channels appear to be weaker, yet the IRD signal level meter registers higher than on other channels.

TI solutions for the analog IRD. With analog receiving systems, you can tune the video slightly away from the TI carrier frequency to receive the best picture. Try adjusting the video frequency to see if the picture improves even if the signal level goes down. On most receivers and IRDs you can store the best setting into memory. You also should suspect microwave interference if sparklies only affect reception of certain satellites and channels; there may be a tower off in that direction. If your receiver does not have a TI filter or this setting does not seem to do enough to clear up the interference, you can use an external "notch" filter to selectively remove the portions of the receiver's IF bandwidth where TI normally will appear. This will remove the relatively narrow telephone microwave carriers from the affected transponder bandwidth, while passing the majority of the satellite signal.

Check the receiver's back panel to determine whether it can accept an "outboard" filter. If the receiver has a short jumper cable linking the input and output ports of its second IF, it can accept a filter.

The frequency of the filter you purchase must match the IF loop-through frequency scheme of the receiver (70, 140, 510 MHz, etc.). Special bandpass filters also are available that can be used to reject interference. Several U.S. companies offer a complete line of filters that can interface with the 2nd IF loop-through frequencies of almost any receiver.

You also should be aware that the presence of TI could prevent the IRD from locking onto one or more encrypted channels. The use of a notch filter is the preferred way to handle this situation. Pass-band filters generally are ineffective for removing TI from encrypted channels because they remove so much of the incoming signal that the decoder will not lock onto the service.

Many analog satellite receivers and IRDs come with built-in TI filters that you can program into memory on a channel-by-channel basis. When using a TI filter, be sure to adjust the video fine tune control for the affected channel until the best reception is obtained.

If your decoder will only intermittently lock onto one or more of the encrypted channels, the channel setting of the receiver may be slightly off center frequency. In this case, use the receiver's video

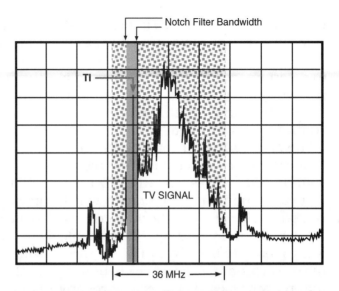

Fig. 10-6. This spectral display of an analog satellite TV signal shows how the notch filter can be used to eliminate an interfering carrier while passing most of the desired signal.

Fig. 10-7. This bandpass filter can be used to remove mild to moderate out-of-band interference. Rejection is 20 dB at 850 MHz and 30 dB at 1550 MHz. (Courtesy Microwave Filter Company.)

Fig. 10-8. This notch filter for commercial satellite TV applications has a passband of 950 to 1450 MHz and can provide TI rejection by up to 40 dB. (Courtesy Communications & Energy Corp.)

fine tune control to adjust the frequency setting so the unit will lock onto and authorize the encrypted service reliably.

TI solutions for the digital IRD. Notch filters usually cannot be used to eliminate terrestrial interference problems for digital DTH receiving systems. Digital satellite TV transmissions typically carry multiple TV and audio services plus conditional access data that is essential for IRD management of the data stream. The essential digital data is spread throughout the entire bandwidth of the transponder in a way that changes continuously. Since a notch filter would eliminate some of the essential data along with the interfering carrier, the IRD would not receive all the information that it needs to function and would therefore stop dead in its tracks.

The most effective way to eliminate terrestrial interference from digital receiving systems, or from analog receiving systems that are "swamped" by a strong TI carrier or carriers, is to block it before it can reach the LNF. It may be possible to move the dish to another location that is shielded from the TI by your house or another building. A solid metal fence or screen, positioned between the microwave source and the

Fig. 10-9. This waveguide filter is installed between the feed and LNB to reject TI caused by marine or airport radar systems. (Courtesy Microwave Filter Company.)

antenna, is another effective method. Make sure the fence or screen is well grounded. In the case of out-of-band interference, special microwave filters are available which can be attached ahead of your LNB.

Preventative maintenance procedures. Freezing temperatures may occur during the winter months in many locales. If you live in a cold weather location, make sure that you take a critical second look at the entire system each year before winter arrives. A few simple preventive maintenance procedures can keep your satellite system from going on the blink.

If you live in an extremely cold climate, try wrapping several feet of heat tape around the actuator's motor and gear box, and then cover it with foam insulation and a plastic bag to prevent the tape and insulation from getting wet. Once the snow begins to fall, periodically brush off the snow before it can accumulate as ice on the antenna surface.

Large-dish C-band systems usually have a plastic cover for the LNB/feedhorn assembly that offers a measure of protection from the elements. Take a moment to inspect the cover to ensure that moisture isn't gaining entry through a crack or seam. Also make sure that the feedhorn's plastic cap is tightly in place. This will prevent moisture from blowing into the throat of the feed where it could freeze and thereby prevent the feedhorn pick-up probe from freely rotating.

INSTRUMENTS FOR SATELLITE WORK

Correct alignment of the receiving antenna and feed-horn polarization is essential if maximum system performance is to be obtained. What's more, the achievement of the highest possible signal levels will provide a measure of margin above the receiver threshold that will help to counteract any Ku-band signal degradation due to rain fade. The fastest and most reliable method of achieving optimum results during the installation procedure—or whenever attempting to locate faulty system components during troubleshooting—is to use an instrument to actually measure the signal strength or "carrier to noise" (C/N) ratio of the incoming satellite signals.

Instruments that measure the signal level of the satellite receiving system vary greatly in price and function. Some of these instruments will not only measure the block IF signal output of the LNF (950 to 1450 MHz or higher) but also the signal levels of terrestrial radio transmissions from about 40 to 860 MHz. These devices can therefore be used for a variety of signal measurement purposes, including but not limited to satellite applications. The following summary explains the applications and benefits of the different types of instruments that currently are available.

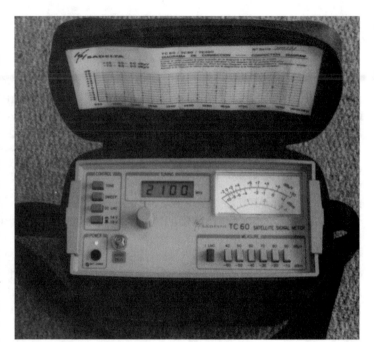

Figure 10-10. The Sadelta T60 satellite signal meter.

rently are available.

Signal meters. One low-cost basic type of signal level measurement device uses a common multi-meter to measure the Automatic Gain Control (AGC) voltage supplied by the satellite receiver, which varies linearly in relation to the incoming satellite signal level. Unfortunately, this indirect method of signal measurement is not highly accurate and many receivers do not provide an easy way for technicians to tap into the AGC voltage. One benefit of the AGC measurement method is that the level of

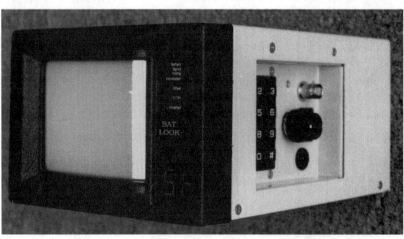

Figure 10-11. A low-cost portable spectrum analyzer.

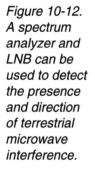

Figure 10-12. A spectrum analyzer and LNB can be used to detect the presence and direction of terrestrial microwave interference.

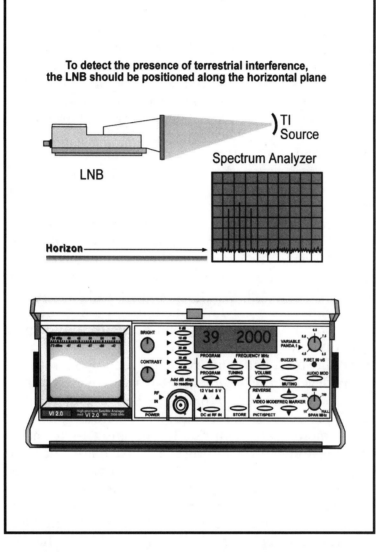

To detect the presence of terrestrial interference, the LNB should be positioned along the horizontal plane

each satellite TV channel can be checked individually. Because the AGC voltage readings are not calibrated, however, there is no way to accurately measure the signal strength and carrier to noise (C/N) ratio of the incoming signals.

A small, lightweight and inexpensive satellite signal meter is a more practical device that should be part of every satellite technician's tool kit. This satellite signal meter can be inserted easily into the coaxial cable linking the LNB and receiver. The same voltage the receiver sends up the cable to power the LNB also provides power to the SF90. No battery or other power source is required to operate the instrument. One disadvantage of this type of signal meter is that it measures all of the signals coming from the satellite at once, so it is of little assistance in making polarization adjustments. The meter readout also is not calibrated, so measurement of actual signal level system C/N is not possible.

More advanced instruments are available that can obtain a much wider range of signal measurements. This type of instrument has an internal power supply so that the antenna and feedhorn can be precisely aligned without a receiver in sight. Moreover, it is fully tunable from 950 - 2050 MHz, so the instrument can be adjusted to zero in on any individual satellite signal of interest. A sweep function allows fast location of any satellite transmission from a given orbital slot. Then, once the instrument is tuned to a particular IF frequency, it will read the actual signal level in either dBm or dBµV. The TC60 also can be used to make accurate system C/N measurements, or even to analyze SMATV system performance. The unit's LNF supply can be switched between 14 and 18 volts, which facilitates the precise adjustment of feedhorn polarization. Technicians can also measure the LNF current to quickly identify an excessive current draw from a faulty unit.

The Spectrum Analyzer. Spectrum analyzers often are used to visually display the incoming satellite signals. The pass-band of the spectrum analyzer can be adjusted to display any portion of the satellite band of interest, from a single satellite signal all the way to the entire frequency band transmitted by any given satellite. Signals appear on the analyzer's screen as "spikes" which rise up from the background noise, or "hash line," displayed at the bottom of the screen. Low-cost spectrum analyzers provide a relative, non-calibrated measurement of signal level, while higher-priced units can be calibrated to an accuracy level of ±1 dB.

Some of the portable, battery-operated spectrum analyzers even contain a satellite TV receiver. The display screens for these units can therefore be switched to either display signal levels or an analog TV picture.

The lower-cost units provide an uncalibrated spectral display with monitor function, an LNB power source, and a polarization switching voltage, all contained in a small hand-held case. Also provided: an audio tone that varies in pitch in direct relation to the level of the incoming signal so that the technician can tweak the antenna without actually needing to see the screen.

The more expensive units feature a calibrated spectral display with an accuracy of better than ±1.5 dB. They may also offer a comprehensive range of auxiliary functions, which includes the display of video sync, pulse and color

burst, as well as a C/N signal level readout. Features include a 14/18-volt switchable LNB power supply and a band inversion switch for the correct display of C-band satellite signals. An optional Calibrated Noise Generator also is available which can be used to measure the frequency response of cables, filters, amplifiers and other system components. Reflectometer measurements, which indicate the distance in a cable to a short or open circuit, also can be made with this device by using a BNC "T" connector and a special calibrated slide rule, both provided with the unit.

All of the above mentioned instruments are analog measurement devices. The new digital hand-held instruments now available have deliberately been excluded from consideration here because the digital nature of these devices limits their ability to faithfully display the incoming signals. First of all, these digital instruments retrace or "refresh" their screens at a relatively slow rate when compared to their analog counterparts. Analog satellite TV transmissions are continually changing in terms of intensity and spectral density. The technician, therefore, may not actually see the true peak value of the signal, or may have difficulty in seeing a minor improvement or degradation in signal levels during the adjustment of the antenna or feedhorn. Furthermore, these digital instruments use a step-tracking tuning that keeps signals that may fall between the steps from being accurately represented on the display screen. This can be a limiting factor for technicians who are attempting to track down interference problems.

HOW TO RECEIVE INTERNATIONAL SATELLITE TV SIGNALS

For those who live beyond the reach of the U.S. and Canadian domsats, there are numerous other regional and international satellites that carry full-time TV program services. What's more, several new international satellites are scheduled for launch between now and the end of the decade. The ongoing satellite explosion that is occurring in Latin America virtually guarantees that satellite TV services of one kind or another are now available at almost every location throughout the Americas.

Reception of these satellite signals is also of great interest to thousands of North American satellite TV enthusiasts who enjoy nothing more than hunting down exotic signals from foreign lands. Be advised, however, that international satellite reception may require specialized equipment and expertise. The following chapter will present the available international, regional and foreign domestic satellite systems and provide you with the information that you will need to succeed in receiving these satellites at your location.

Due to the rapidly changing nature of the international satellite environment, this chapter will not attempt to provide a satellite-by-satellite list of the available TV and radio services. For the most up to date information, I recommend that you visit the web site addresses that are provided at the end of each satellite system entry, or use one of the many available satellite list services now on the Internet, such as the one at http://www.satcodx.com on the worldwide web.

Readers should also be aware that television accounts for just a fraction of the telecommunication traffic handled by most of these satellite systems. The traffic on any regional or international satellite will invariably be a combination of many different types of signals, from TV news, sports and entertainment to long distance telephone conversations and private business data networks. Part of the challenge is learning the capabilities of each system so that you will know where to look and how to tune in the signals that you encounter. Happy hunting!

Fig.11-1. INTELSAT VII satellite structural layout.

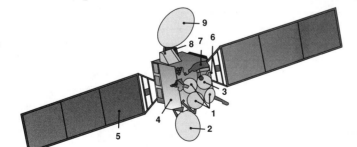

1. Steerable Ku-band spot beam antenna.
2. C-band receive hemi/zone antenna.
3. Steerable C-band spot beam antenna.
4. Thermal radiator.
5. Solar array.
6. C-band global beam.
7. C-band transmit hemi/zone feed array.
8. Hinges.
9. C-band transmit hemi/zone antenna.

INTERNATIONAL TV DISTRIBUTION SATELLITES COVERING THE AMERICAS

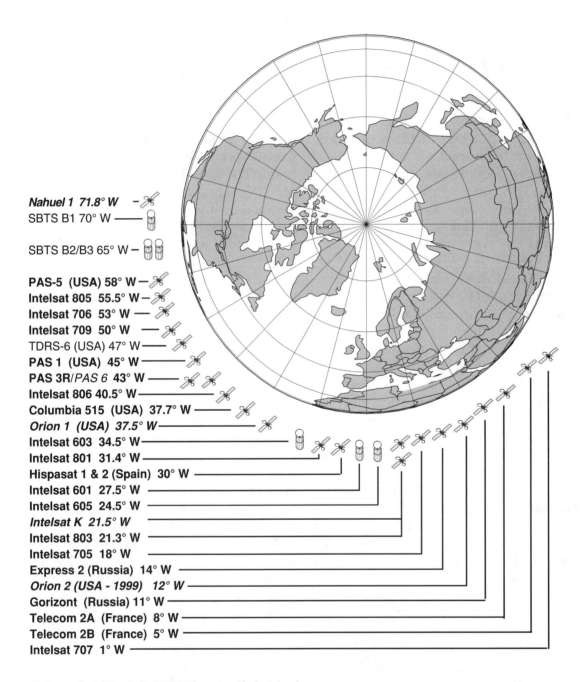

Nahuel 1 71.8° W
SBTS B1 70° W
SBTS B2/B3 65° W
PAS-5 (USA) 58° W
Intelsat 805 55.5° W
Intelsat 706 53° W
Intelsat 709 50° W
TDRS-6 (USA) 47° W
PAS 1 (USA) 45° W
PAS 3R/*PAS 6* **43° W**
Intelsat 806 40.5° W
Columbia 515 (USA) 37.7° W
Orion 1 (USA) 37.5° W
Intelsat 603 34.5° W
Intelsat 801 31.4° W
Hispasat 1 & 2 (Spain) 30° W
Intelsat 601 27.5° W
Intelsat 605 24.5° W
Intelsat K 21.5° W
Intelsat 803 21.3° W
Intelsat 705 18° W
Express 2 (Russia) 14° W
Orion 2 (USA - 1999) 12° W
Gorizont (Russia) 11° W
Telecom 2A (France) 8° W
Telecom 2B (France) 5° W
Intelsat 707 1° W

Fig. 11-2. International TV distribution satellites serving the Americas. On March 31, 1998, INTELSAT announced that it is forming a spin-off company called New Skies Satellites that will take over the operation of the INTELSAT 803 (21.5° W), INTELSAT K (21.5° W) and INTELSAT 806 (40.5° W) satellites in the AOR. The transfer of these assets will allow the new company to seek public funding. It will not affect the technical characteristics or performance parameters of these satellites in any way.

Note: Satellites in italic type face are Ku-band only.
       Satellites in plain type face are C-band only.
       Satellites in bold type face are dual C/Ku-band.

THE INTELSAT SATELLITE SYSTEM

INTELSAT is the largest commercial satellite service provider in the world, with more satellites in operation than any other organization around the globe. Founded in 1964, INTELSAT was the first telecommunication organization to provide global satellite coverage and connectivity. Today's INTELSAT global satellite system brings television, telephone, and data distribution services to billions of people in more than 200 nations, territories, and dependencies on every continent.

INTELSAT is an international, non-profit cooperative consisting of more than 140 member nations. The INTELSAT owners contribute capital and receive a return on their investment in proportion to their actual use of the system. Any nation may use the INTELSAT system, whether or not it is a member, in exchange for payment for the capacity that it uses, with cost based on the signal bandwidth and the type of capacity that is used.

THE INTELSAT SATELLITE FLEET

The INTELSAT satellite fleet consists of over 20 telecommunication spacecraft in geostationary orbit: the INTELSAT V/V-A, INTELSAT VI, INTELSAT K, INTELSAT VII/VII-A and INTELSAT VIII/VIII-A series. Moreover, INTELSAT already is working on its newest generation of INTELSAT spacecraft, the INTELSAT IX series, four models of which were in production at the time of writing.

The INTELSAT V/V-A Series. As of April 1998, INTELSAT had seven INTELSAT V/V-A series spacecraft in service, with four of these satellites operating in the Atlantic Ocean Region (AOR). Manufactured during the 1980s, the INTELSAT V spacecraft were the first INTELSAT satellites capable of transmitting and receiving signals in both the C and Ku satellite frequency bands. They were also the first spacecraft to use the spatial separation between coverage beams AND employ two orthogonal senses of polarization so that portions of the available satellite frequency spectrums could be reused four times for maximum efficient use of each INTELSAT orbital resource. All of the remaining INTELSAT V satellites in orbit have exceeded the V series' nominal mission lifetime of seven years. The extension of mission lifetime has been achieved through the use of inclined orbit techniques that will be discussed later in this chapter.

The INTELSAT VI Series. As of April 1998, there were five dual-band INTELSAT VI series satellites in operation, with three of these satellites in service in the Atlantic Ocean Region. At the time of their launch in the early 1990s, the INTELSAT VI series satellites were the largest commercial spacecraft ever built. The INTELSAT VI satellites were specifically designed to handle a type of broadband data traffic called satellite-switched time division multiple access (SS/TDMA). This innovative technology makes it possible for each INTELSAT VI satellite to dynamically interconnect six defined C-band coverage areas. In another first, the INTELSAT VII series includes two southeast and southwest zone beams in addition to the northeast and northwest zone beams carried by the INTELSAT V series satellites.

The INTELSAT VII/VII-A Series. As of April 1998, there were eight dual-band INTELSAT VII/VII-A series satellites in operation, with four of these satellites in service in the Atlantic Ocean Region. Although the initial design of the INTELSAT VII series focused on the special requirements of the Pacific Ocean

Region, all of these satellites have the versatility to operate effectively at any INTELSAT orbital location around the world. Improving previous generations of INTELSAT spacecraft, the INTELSAT VII/VII-A series features three independently-steerable, medium-power, C-band AND Ku-band spot beams that can be individually steered to cover any area visible from the satellite's orbital location on an as needed basis.

The INTELSAT K Satellite. INTELSAT K is a dedicated Ku-band-only spacecraft launched in 1992 to an orbital assignment over the Atlantic Ocean Region (AOR) to meet the transatlantic requirements of various international broadcasting organizations. The INTELSAT K satellite was the first INTELSAT spacecraft to provide Ku-band downlink services into Latin America to antennas as small as 1.2 meters in diameter. The spacecraft was also used to deliver the first ever transatlantic DTH TV broadcasting service.

The INTELSAT VIII Series. As of April, 1998, a total of five dual-band INTELSAT VIII and VIII/A satellites were in operation, with two providing services in the Atlantic Ocean Region (AOR) and a third on its way to the AOR region to begin operations later in the year. The INTELSAT VIII/VIII-A series features the highest C-band power level ever for an INTELSAT satellite. One of the INTELSAT VIII-A satellites (806) has been specifically designed to provide cable TV distribution services within a wide area stretching from the Americas all the way to Western Europe.

The INTELSAT IX Series. INTELSAT currently is developing a series of dual-band Follow On Satellites (FOS) called the INTELSAT IX series, the first of which is scheduled for launch in the year 2000. The INTELSAT IX series will feature the largest satellite capacity in the INTELSAT fleet, with expanded high-power coverage beams and full interconnection among beams. Due to these and other innovations, the INTELSAT IX series will be able to deliver a wide range of services to small aperture Earth stations, including high-speed Internet links, broadcast and cable TV, and digital DTH signal distribution. The first two flight models of this series are set to replace two INTELSAT VI satellites in the Indian Ocean Region (IOR) that will be approaching the end of their nominal mission lifetimes at the turn of the millennium.

INTELSAT COVERAGE BEAMS

With the exception of the INTELSAT K satellite, all of the above-mentioned INTELSAT satellites carry three distinct types of C-band coverage beam antennas: global, hemispheric and zone. What's more, some of the later INTELSAT V satellites, as well as all of the INTELSAT VII spacecraft, also carry high-powered steerable C-band spot beams. The type of beam for many INTELSAT transponders can be selected by ground command for routing TV programs and other traffic to their intended destinations.

Global Beams. The global beam covers the largest area of any available beam, the 42.4 percent of the Earth's surface that is actually visible from any given geostationary orbital location. INTELSAT satellites have used global beams ever since the launch of the very first INTELSAT "Early Bird" satellite back in 1965. These global beams are widely used to relay TV news, sports and entertainment programs from one side of the Atlantic, Indian or Pacific Oceans to the other. These global beam feeds range from short news items of three minutes or less to live coverage of sports competitions or other major events. International programmers may also use

Fig. 11-3, Fig. 11-4, Fig.11-5 & Fig.11-6. INTELSAT 601 global, hemi, zone and spot beam coverage from 34.5° west (325.5° east) longitude. Numerous other satellite coverage maps and computational tables appear in Appendix A.

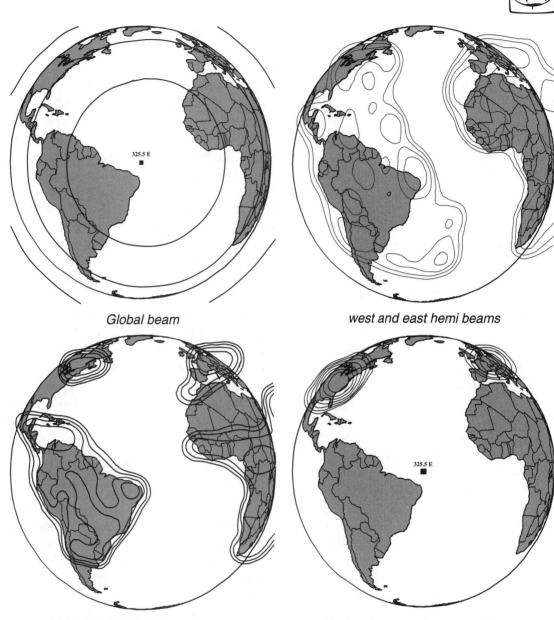

Global beam

west and east hemi beams

NW, SW, NE & SE zone beams

Ku-band west and east spot beams

INTELSAT global beams because the coverage area is so widespread.

Because the satellite signal is projected over such a vast area, however, large dish antennas usually are required in order to receive the relatively weak signals that are produced. The exact size of antenna required will vary depending on how much power the particular INTELSAT satellite puts into the global beam. Antenna size typically ranges from five to ten meters in diameter.

Hemispheric Beams. Several countries worldwide use INTELSAT hemispheric beam transponders to transmit television programming. The hemispheric beam coverage area is reduced to less than half of the global beam area, covering approximately 20 percent of the Earth's surface to the east or west of the satellite's orbital location. This allows smaller and less expensive receiving antennas to be used down on the ground. Satellite antennas in the three to six meter range typically are required to produce a high quality signal when receiving these hemispheric beam transmissions.

Zone Beams. The zone beam transmits its signal into an area less than half the size of a hemispheric beam. This doubles the signal intensity, allowing dishes less than four meters in diameter to perform effectively in many cases. INTELSAT V satellites feature two zone (northeast and northwest) beams which reach locations above the Earth's equator, while INTELSAT VI and VII series spacecraft provide a total of four (northeast, southeast, northwest and southwest) zone beams.

Spot Beams. The later INTELSAT V satellites (INTELSAT 510 through 515) and all INTELSAT VII satellites carry a steerable C-band spot beam that can be commanded to point at any location that is visible from the satellite's orbital position. All INTELSAT satellites, including INTELSAT K, also carry two or more Ku-band spot beams. While INTELSAT K's spot beams are fixed for coverage of Europe, North and South America, the remaining INTELSAT spacecraft have steerable Ku-band spot beams that can be re-pointed to cover any location that is visible from the satellite's assigned orbital location. All of these spot beams, which are tightly focused to provide coverage of relatively small areas of the Earth's surface, can be used to deliver TV services into small-aperture antennas that are located within their respective coverage areas.

INTELSAT POLARIZATION STANDARDS

All INTELSAT C-band satellite transmissions use circular polarization. Instead of positioning the microwave energy in a vertical or horizontal plane, circular polarization is transmitted in a helical pattern, with the wave front rotating either clockwise (right-hand circular) or counterclockwise (left-hand circular) as viewed from the satellite.

INTELSAT satellites are capable of simultaneously transmitting signals using both right-hand and left-hand circular polarization into any given satellite coverage area. This allows the limited C-band frequency range to be re-used without causing interference.

For example, INTELSAT satellites use only left-hand circular polarization for their zone beams, while the global and hemispheric beams can use either left-hand or right-hand circular polarization in most cases. This cross-polarization between beams allows simultaneous use of right-hand circular polarization hemispheric or global and left-hand circular polarization zone beams within the same coverage areas.

INTELSAT 806 FOR LATIN AMERICA

On February 27, 1998, the INTELSAT 806 communications satellite was successfully launched to an orbital assignment of 40.5 degrees west longitude. The new spacecraft will deliver the largest video cable community on any single INTELSAT satellite, including the delivery of an impressive channel lineup of over seventy of the most popular regional and international video channels to over 2,000 cable TV head ends throughout the Americas.

To create this instant community, many programmers from other INTELSAT satellites, as well as new customers, will migrate to INTELSAT 806 to take advantage of the strength of the new digital neighborhood's established audience of over 20 million viewers in Latin America. International programmers also will gain coverage of

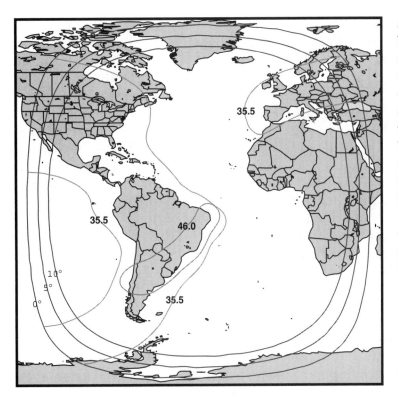

Fig. 11-7. INTELSAT 806 C-band and Ku-band coverage zones from 40.5° west longitude. Peak EIRP values are +4 dBW higher than the beam-edge values shown.

at no extra charge. C-band antennas as small as 1.8 meters in diameter will be able to receive signals from the satellite at any location within the C-band beam's primary coverage area. For further information on the INTELSAT satellite system, visit the INTELSAT web site at http://www.intelsat.int on the worldwide web.

THE PANAMSAT SATELLITE SYSTEM

In May of 1997, a new PanAmSat Corporation was formed through the merger of PanAmSat and Galaxy Satellite Services, a division of Hughes Communications. The PanAmSat global satellite network currently consists of seventeen satellites. As of April 1998, there were six PAS satellites in operation, including four (PAS-1, PAS-3R, PAS-5 & PAS-6) over

Europe at the same time due to the satellite's unique coverage beam, which will provide simultaneous connectivity to Latin America, the U.S. and Europe. For example, services uplinked in Europe will also be simultaneously downlinked in South and North America

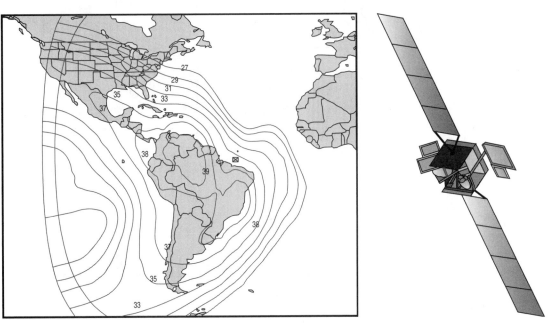

Fig. 11-8. One of the many coverage beams provided by the PAS-3R satellite at 43° west longitude. See Appendix A for additional coverage maps for all the available satellites as well as the EIRP to dish size conversion charts.

the Atlantic Ocean Region (AOR). Five additional PAS satellites—two (PAS-1R and PAS-6B) for the AOR, two for the IOR and one for the POR—are scheduled for launch within the next two years. PanAmSat also operates the Galaxy and SBS satellite fleets covering North America, with Galaxy III-R and Galaxy VIII-I also providing digital DTH services for Latin America. A new Galaxy XI satellite is also scheduled for launch in the near future that will have the capability to deliver high-power digital DTH signals to the Latin American market. Like the previously described INTELSAT satellite system, TV traffic accounts for just a fraction of the telecommunication traffic on the PanAmSat system. For further information on the PAS, Galaxy and SBS satellites, visit the PanAmSat web site at http:// www.panamsat.com.

THE INTERSPUTNIK SATELLITE SYSTEM

Established in 1971 by nine member countries, INTERSPUTNIK is an international organization with twenty-three member states, several of which are either members of the Commonwealth of Independent States or Eastern European nations. The INTERSPUTNIK system consists of a space segment that includes communications satellites either owned by the organization or leased from member nations.

INTERSPUTNIK currently leases Gorizont, Express and Gals satellites from the Russian Federation. At year-end 1997, INTERSPUTNIK was using a total of 31 transponders on seven geostationary satellites located in the orbital arc from 14 degrees west longitude to 142.5 degrees east longitude, two of which can be received from locations in the Americas.

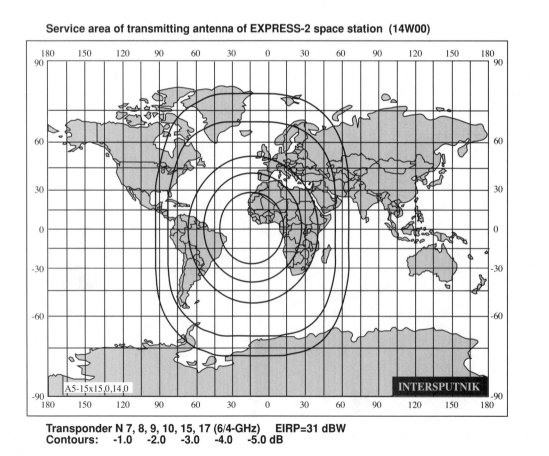

Service area of transmitting antenna of EXPRESS-2 space station (14W00)

A5-15x15,0,14,0

INTERSPUTNIK

Transponder N 7, 8, 9, 10, 15, 17 (6/4-GHz)   EIRP=31 dBW
Contours: -1.0   -2.0   -3.0   -4.0   -5.0 dB

Fig. 11-9. Express 2 global beam coverage from 14° west longitude.

In 1997, INTERSPUTNIK entered into the Lockheed Martin INTERSPUTNIK (LMI) Joint Venture, with INTERSPUTNIK and Lockheed Martin of the U.S. serving as partners in the development of a new state-of-the-art satellite system. Under the new Joint Venture, Lockheed Martin will raise financing and provide logistic support to LMI, while INTERSPUTNIK will be the system operator of the new satellites. The first geostationary LMI satellite is scheduled for launch to the Indian Ocean Region (IOR) in late 1998. Four additional LMI satellites are in the planning stages for other orbital locations around the world.

THE INTERSPUTNIK SATELLITE FLEET

Like the INTELSAT satellites, the Russian Gorizont and Express spacecraft provide four types of beam coverage: global, hemispheric, zone, and spot.

Russian ground control stations can select between global, hemispheric, zone, and fixed spot beam antennas on designated transponders for maximum versatility. While INTELSAT's hemispheric and zone beams cover the eastern and western portions of the globe, the Gorizont hemispheric and zone beams are biased to maximize coverage of locations north of the Earth's equator.

Gorizont Satellites. Gorizont satellites carry six C-band transponders operating in the 3.65 to 4.0 GHz frequency spectrum and one Ku-band transponder operating in the 11.5 to 11.6 GHz frequency range. Gorizont satellites usually only carry enough stationkeeping fuel to allow for three years of geostationary operations plus several years of extended life through the use of a fuel-conserving technique called an "inclined orbit." The Gorizont satellite stationed at 11 degrees west longitude is currently the only one that can be received from locations in the Americas. All ground stations receiving signals from an inclined orbit satellite must track the movements of the spacecraft over each 24-hour period. Information on how to track Gorizont and other inclined orbit satellites is provided later in this chapter.

Express Satellites. In 1995, The Russian Space Agency deployed the first in a new series of "Express" communications satellites to an orbital assignment

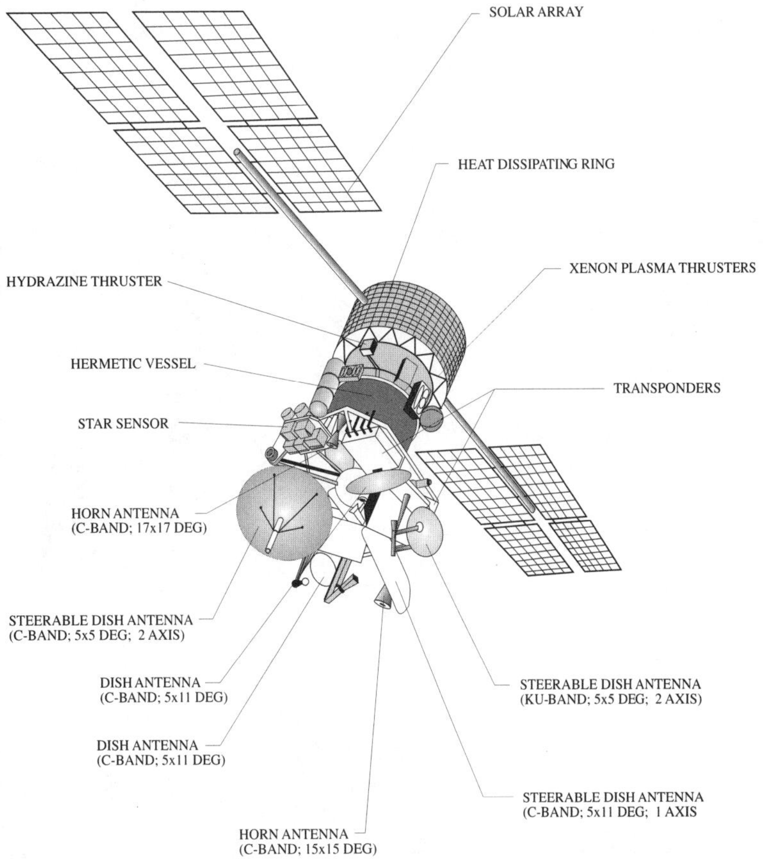

Fig. 11-10. Structural layout of the Express 2 satellite.

SOLAR ARRAY

HEAT DISSIPATING RING

XENON PLASMA THRUSTERS

HYDRAZINE THRUSTER

HERMETIC VESSEL

STAR SENSOR

TRANSPONDERS

HORN ANTENNA
(C-BAND; 17x17 DEG)

STEERABLE DISH ANTENNA
(C-BAND; 5x5 DEG; 2 AXIS)

DISH ANTENNA
(C-BAND; 5x11 DEG)

STEERABLE DISH ANTENNA
(KU-BAND; 5x5 DEG; 2 AXIS)

DISH ANTENNA
(C-BAND; 5x11 DEG)

STEERABLE DISH ANTENNA
(C-BAND; 5x11 DEG; 1 AXIS)

HORN ANTENNA
(C-BAND; 15x15 DEG)

over the Atlantic Ocean Region at 14 degrees west longitude. Each Express satellite carries ten C-band and two Ku-band transponders. On C-band, the Express satellites offer fully steerable zone and spot beams for coverage of any land mass visible from the satellite's assigned orbital location over the Earth's equator.

Future Russian Satellites. Russia is now in the process of acquiring modern replacements for its antiquated fleet of Gorizont satellites. A new class of "Yamal" satellites is now under construction to replace several Gorizont-class spacecraft. For further information on the Gorizont, Express and LMI satellites, visit the INTERSPUTNIK web site at http://www.intersputnik.com on the worldwide web.

THE COLUMBIA SATELLITE SYSTEM

Columbia Communications Corporation is a U.S. satellite telecommunica-tions company providing domestic and international voice, data and video services covering a geographic area stretching from the Asian Pacific Rim, throughout the Americas, to Western Europe, Eastern Europe, and Africa.

In January 1992, Columbia initiated commercial services using commercial C-Band transponder payloads on two of NASA's TDRS (Tracking and Data Relay Satellite) spacecraft, which also supply telecommunication support for space shuttle missions and other space activities. The coverage beams of the TDRS satellites over the Atlantic and Pacific Oceans overlap in the central United States. This allows Columbia to provide connectivity between Asia, North America and Europe.

As of April 1998, Columbia controlled a total of fifty-two C-band transponders on four different satellites, including three in the Atlantic Ocean Region (AOR) at 27.5 degrees west longitude (capacity

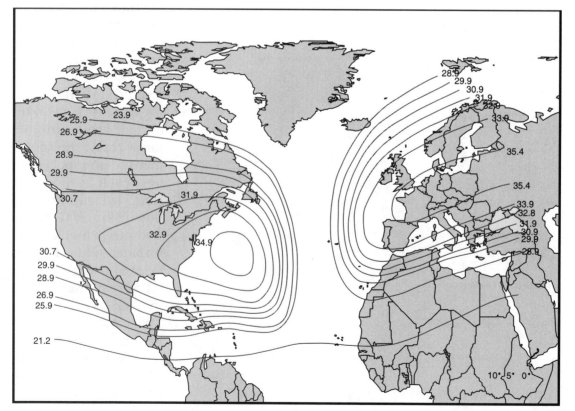

Fig. 11-11. TDRS-6 C-band coverage beam from 47° west longitude.

on INTELSAT 605), 37.7 degrees west longitude (COLUMBIA-515) and 47 degrees west longitude (TDRS-6).

Columbia has announced plans to begin replacing its existing capacity by the year 2000 with a new series of satellites to be manufactured by TRW. At least one of these new spacecraft will cover the Americas in their entirety. For further information on the Columbia satellite system, visit their Internet site at http://www.tdrss.com on the worldwide web.

THE ORION SATELLITE SYSTEM

Orion Network Systems, Inc. is an international satellite service provider that delivers high-speed Internet connectivity and private multimedia services directly to multinational businesses equipped with small-aperture receiving antennas. At the time of writing, the Orion system consisted of a single Ku-band satellite located at 37.5 degrees west longitude.

Launched in November of 1994, Orion 1 is a Ku-band-only satellite with spot beam coverage of Western Europe and North America. Operator Orion Network Systems—which was acquired by Loral Corporation in 1997—intends to launch two additional satellites before the end of the decade. Orion 2 will provide spot beam coverage of Western Europe, Latin America and the eastern United States from an orbital assignment of 12 degrees west longitude. The dual-band Orion 3 satellite will provide coverage of Hawaii and eastern Asia from an orbital assignment of 139 degrees east longitude. For further information on the Orion satellite system, visit the Orion Internet site at http://www.orionnetworks.net on the worldwide web.

THE SBTS SATELLITE SYSTEM

Brazil's SBTS (Sistema Brasileiro de Telecomunicaoes por Satelite) satellite system is owned and operated by EMBRATEL, the country's government-owned telecommunications service provider and INTELSAT member. The first generation of SBTS satel-

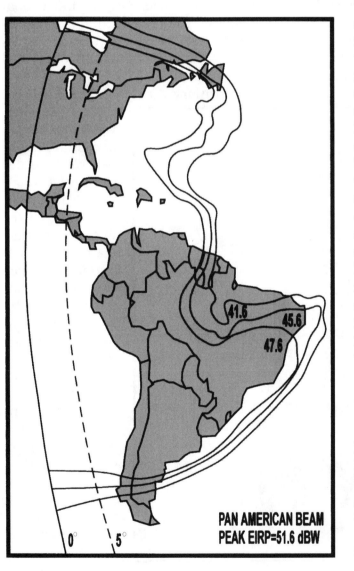

Fig. 11-12. Orion-2 Ku-band coverage from 12° west longitude. Orion-2 is scheduled for launch in 1999.

PAN AMERICAN BEAM
PEAK EIRP=51.6 dBW

lites, Brasilsat A1 and A2, were launched in 1982. Both of these C-band satellites have since been replaced by the Brasilsat B1 and B2 spacecraft presently located at 65 and 70 degrees west longitude. Manufactured by Hughes Space and Communications, Brasilsat B1 and B2 each carry 28 C-band transponders that downlink within the 3.65 to 4.2 GHz frequency spectrum. The EIRP for the national and regional (covering major urban areas) beams varies from 33 to 41 dBW. The coverage zone of both beams is restricted to Brazil and the adjacent nations. Reception from locations in North America is not possible. Each satellite also carries one X-band transponder for military communications. A third satellite, Brasilsat 3A, was successfully launched in early 1998. The satellite is identical in design to Brasilsat

B1 and B2, except that B3 does not carry an X-band satellite transponder. For further information on the SBTS satellite system, visit the EMBRATEL web site at http://www.embratel.com.br on the worldwide web.

THE NAHUEL SATELLITE SYSYTEM

The Nahuel Satellite System is owned and operated by NahuelSat S.A., a privately owned company that was awarded a 24-year concession by the government of Argentina. The company's major shareholders include the following: Daimler Benz Aerospace, Aerospatiale, Alenia Aerospazio, GE Capital Global Satellites, the International Finance Corporation, Publicom S.A., BISA Group, Banco Provincia Group, and Antel of Uruguay.

On January 30, 1997, the Ku-band Nahuel 1 satellite was launched to an orbital assignment of 71.8 degrees west longitude. The spacecraft carries three different coverage beams, of which one encompasses a vast region that extends from the Antarctic and Tierra del Fuego to the southern U.S. A second Ku-band beam covers all of Argentina, Chile, Uruguay, Paraguay and parts of Brazil and Bolivia, while the third beam supplies regional coverage of all Latin Ameri-

Fig. 11-13. Brasilsat B structural layout. Coverage maps for Brasilsat B1 and B2 appear in Appendix A.

Diameter 3.65 m (12 ft)

Height stowed 3.43 m (11 ft 3 in)

Height deployed 8.3 m (27 ft 3 in)

TELEMETRY AND COMMAND ANTENNA

ANTENNA REFLECTOR

DESPUN REPEATER PANEL

ANTENNA FEEDS

FIXED SOLAR PANEL

SUN SENSOR (2)

SOLID STATE POWER AMPLIFIER

THERMAL RADIATOR

BATTERY PACK (4)

RADIAL THRUSTER (2)

EARTH SENSOR (2)

PROPELLANT TANK (4)

NUTATION ACCELEROMETER

DEPLOYABLE SOLAR PANEL

can countries.

The Argentine National Commission of Communications and NahuelSat are currently investigating the possibility of increasing the system's capacity through the launch of a Nahuel 2 satellite to a different orbital location. To that end, the two organizations are involved in coordinating with the ITU the use of the secondary orbital locations of 59, 76, 80 and 85 degrees west longitude. For further information on the NahuelSat satellite system, visit http://www.nahuelsat.com.ar.

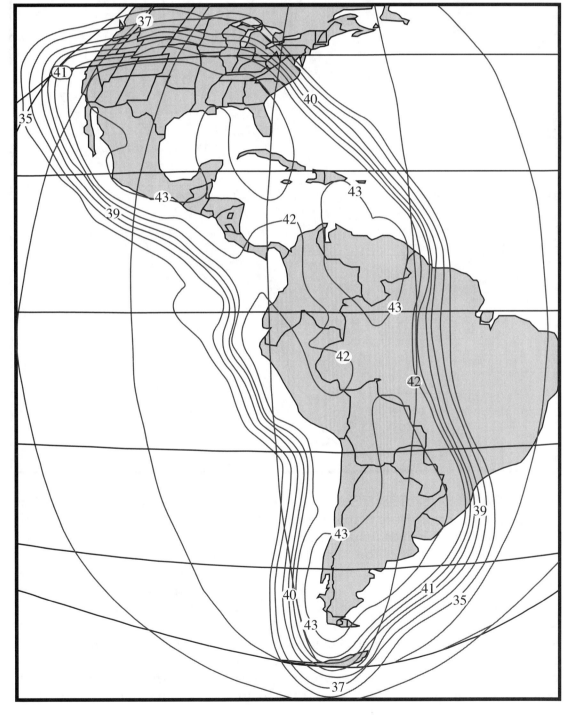

Fig. 11-14. Nahuel broad beam coverage of the Americas from 71.8° west longitude.

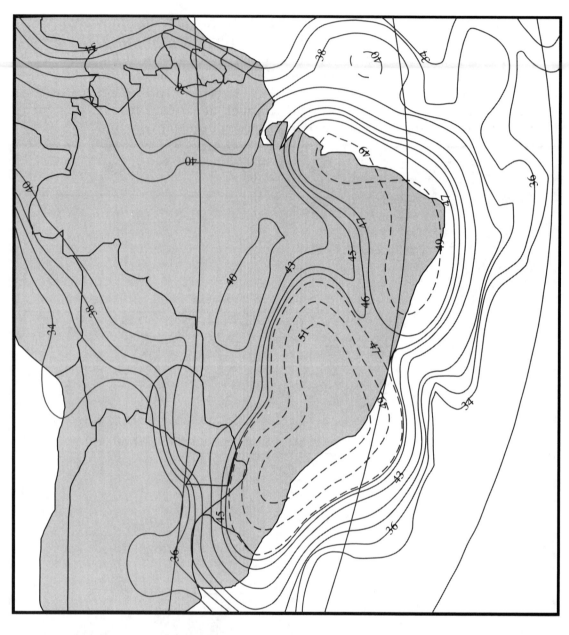

Fig. 11-15. Nahuel-1 regional beam coverage of Brazil from 71.8° west longitude.

THE HISPASAT SATELLITE SYSTEM

Hispasat is the only European system that offers satellite telecommunication services on both sides of the Atlantic. Collocated at 30 degrees west longitude, the Ku-band Hispasat 1A and Hispasat 1B satellites feature high-power coverage of the Americas, which allow the satellites to transmit analog and digital DTH TV services. Hispasat's broad coverage beam for the Americas, which reaches all the way from Canada to Patagonia, is strong enough to allow antennas from 90 to 120 cm to receive TV programs from the two satellites. The Hispasat Company also intends to launch a third satellite to meet the increasing demands for digital DTH service in both the Spanish and Ibero-American markets. Hispasat 1C is scheduled to become operational by the end of 1999.

The Hispasat broadcasts to the Americas include a four-channel digital bou-

Fig. 11-16. Hispasat 3 Ku-band coverage of the Americas from 30° west longitude. Hispasat 3 is scheduled for launch in 1999.

quet from RadioTelevision Española that is carried on transponder 15 (12.015 GHz, vertical polarization). Reception requires an MPEG-2 DVB-compliant IRD tuned to a symbol rate of 27.5 Msym/sec and an FEC rate of 3/4.

Hispasat also offers an analog DTH service from Television Española via transponder 6 (12.078 GHz, vertical polarization). The main TV audio uses a 6.60 MHz subcarrier, while Spain's Radio Nacional and Radio Exterior use 7.38 and 7.56 MHz audio subcarriers, respectively. For further information on the Hispasat satellite system, visit the Hispasat Internet site at http://www.hispasat.com on the worldwide web.

THE SATMEX SATELLITE SYSTEM

In 1997, the Mexican government established an auction for the sale of 75 percent of the assets of Satélitas Mexicanos (SatMex), the state-owned national satellite operator. In early 1998,

the government announced that a joint venture bid submitted by Loral Corporation and Mexico's Telefonica Autrey, S.A. de C.V. had been selected as the winner. The assets of SatMex consist of four satellites, Solidaridad 1, Solidaridad 2, Morelos II, and the Morelos III satellite now under construction, along with twenty-year concessions for the use of three orbital slots together with renewal rights for an additional twenty-year period.

The Solidaridad Satellites. Mexico's Solidaridad 1 and Solidaridad 2 satellites are based on the Hughes HS601 spacecraft bus. The two satellites provide such services as voice telephony, data communications, television relay, facsimile transmission, business networks, educational TV broadcasts, and a nationwide satellite mobile telephone service.

Each Solidaridad spacecraft carries 18 active C-band transponders with amplifiers ranging from 10 to 16 watts of power that feed into various regional beams. Each Solidaridad satellite also carries sixteen active Ku-band transponders that are equipped with 42.5-watt amplifiers.

The C-band and Ku-band national beams cover all of Mexico as well as the southwestern United States. Moreover, additional Ku-band spot beams target such major U.S. cities as Chicago, Dallas, Houston, Los Angeles, Miami, New York, San Antonio, San Francisco, Tampa, and Washington, DC, while Solidaridad's C-band regional beam coverage includes the Caribbean, Central and South America. Solidaridad 1 and 2 are located at 109.2 and 113 degrees west longitude, respectively.

The Morelos Satellites. Launched back in 1985, Morelos II is a dual-band satellite that is scheduled for retirement in 1998. Its replacement at 116.8

degrees west longitude will be Morelos III, a Hughes HS 601 satellite that features 7,000 watts of payload power, or more than 10 times the power capacity of the satellite of its predecessor.

Morelos III will carry 24 C-band transponders with 36-watt amplifiers as well as 24 high-power Ku-band transponders equipped with 132-watt amplifiers. The Ku-band capacity will be used to deliver digital DTH services into antennas with a diameter of 60cm or smaller. The satellite also provides three separate coverage beams, with one for C-band service and the other two for Ku-band service. Through the use of these three beams, the spacecraft will be able to deliver services to locations throughout the Americas, including countries such as Argentina, Brazil, Chile, Colombia, El Salvador, Peru, Venezuela and the United States, among others.

TELECOM

In addition to relaying TV programming into Western Europe via their Ku-band transponders, the dual-band French Telecom 2 series satellites located at 8 and 5 degrees west longitude also transmit French TV programming to overseas territories in the Americas and Africa using C-band transponders with left-hand circular-polarization. The available programming includes video feeds for RFO France.

RECEPTION REQUIREMENTS

The appropriate dish size for international C-band satellite reception at your location will depend on which beam pattern is being used by the particular satellite or satellites that you wish to view. INTELSAT, TELECOM and STATSIONAR reception also requires the use of special satellite feedhorns, receivers, and antenna tracking devices

in order to obtain a high-quality picture.

Circular Polarization. For the best possible reception of INTELSAT, TELECOM and STATSIONAR satellites, you will need to change your feed polarization from horizontal/vertical (which is used by the Galaxy, GE, Gstar, Spacenet, Telesat, Telstar and PAS satellites) to circular polarization. Instead of positioning the microwave energy in a plane, whether vertical or horizontal, circular polarization is transmitted in a circular pattern that rotates in either a clockwise or a counterclockwise direction. Although your regular feedhorn can still pick up circularly polarized signals, you end up losing half of the available signal. This can mean the difference between a noise-free picture and one that is not tolerably viewable.

Special international feedhorns have been created which can receive both linearly and circularly polarized satellite signals, so that C-band dish owners in the Americas can receive programming from the available international and regional satellites. These products are available from several leading satellite equipment manufacturers.

When your feedhorn skew is adjusted to optimally receive signals using right-hand circular polarization (RHCP), you will be unable to see any left-hand circular polarization (LHCP) signals, and vice versa. This cross-polarization isolation is identical to the isolation between the horizontal and vertical polarizations used by domestic and regional satellites. To maximize your reception of both circular and linear satellite transmissions, follow the instructions provided by the feedhorn manufacturer when installing the feed on your system.

The Receiver IF Frequency Band. All LNBs convert the incoming satellite frequency band from the original frequencies to a lower set of intermediate frequencies. For North American C-band and Ku-band LNBs, the block IF that is produced usually ranges from 950 to 1450 MHz. For specific applications, some Ku-band LNBs may use a wider block IF of 950 to 1950 MHz or higher.

Ku-band LNBs contain a "local" oscillator (LO) circuit that generates a signal that is lower in frequency than the incoming Ku-band signal. The LO frequency beats or "heterodynes" with the incoming satellite spectrum to produce a block IF output signal that is the difference between the original satellite frequency and the local oscillator frequency. This is called the "low side injection" method. For example, the 11.7 to 12.2 GHz incoming satellite frequency band heterodynes with the LNB's LO frequency of 10.75 GHz to produce a block IF of 950 to 1450 MHz. The satellite transponders would appear within the block IF in the same order as they appear in the original satellite band.

C-band LNBs also use a local oscillator to produce the IF band for the receiver, but one with an oscillator frequency that is higher than the incoming C-band satellite frequency spectrum. This is called "high side injection." For example, if the incoming satellite spectrum ranges from 3.7 to 4.2 GHz and the LNB's LO frequency is 5.15 GHz, the resulting difference between the two would be a block IF of 950 to 1450 MHz. With high side injection, the presentation order of the transponders in the IF band is reversed and the video polarity is inverted. Dual-band receivers and IRDs are designed to compensate for the reversal of channel order and the inversion of the video when switching from Ku-band to C-band reception modes. When using a spectrum analyzer to examine the block IF output of a C-band LNB, however, you will need to keep in mind that the channel order of the analyzer's display will be inverted.

Fig. 11-17 & Fig. 11-18. Examples of High-Side versus Low-Side frequency injection methods used to produce the LNB block IF output.

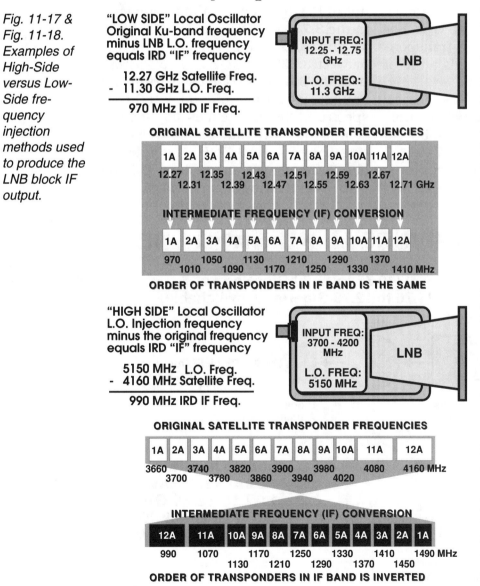

Some spectrum analyzers have an inversion switch that reverses the C-band display back to the correct order.

The C-band satellite transponder charts presented in Appendix B provide the relevant satellite transponder frequencies. Some receivers and IRDs, however, require that the operator enter the IF frequency for each satellite transponder into the unit's memory before tuning to that transponder. Almost all C-band LNBs use an LO frequency of 5.15 GHz. The correct IF frequency can be determined by subtracting the original satellite transponder center frequency from the LO frequency of 5.15 GHz. For example, 5.15 GHz minus a transponder frequency of 3.725 equals an IF of 1.425 GHz, or 1425 MHz.

Broadband "universal" LNBs are now available in Europe and elsewhere overseas that allow a single Ku-band LNB to receive services in both the 10.7 to 11.7 GHz and 11.7 to 12.75 GHz spectrums. The receiver or IRD activates a 22 kHz tone which switches the LNB's local oscillator frequency between two values (for example, 9.75 GHz or 10.6 GHz) to produce a low-band IF output of 950 to 1950 MHz or a high-band IF output of 1100 to 2150 MHz. To effectively use one of these products, you must have an IRD that has an IF input range which matches the IF output range of the universal LNB.

Various C-band frequency schemes are used by the available international and regional satellites mentioned above. The INTELSAT V and VII series satellites, as well as the PAS satellites, transmit to Earth using C-band frequencies in the 3.7 to 4.2 GHz range. INTELSAT VI and Brasilsat B satellites operate between 3.625 to 4.2 GHz, while Gorizont uses the 3.65 to 4.0 GHz spectrum and Express uses the 3.65 and 4.180 GHz spectrum. Newer satellites may soon be coming our way that will be transmitting in the 3.40 to 3.7 GHz "extended C-band" frequency range. Many C-band LNB manufacturers now make products which can receive the new ex-

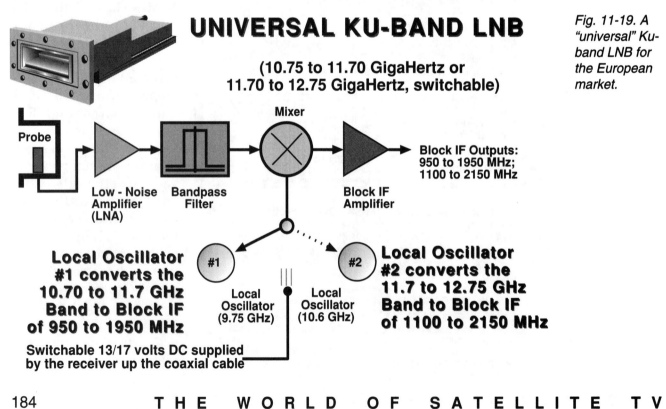

UNIVERSAL KU-BAND LNB

(10.75 to 11.70 GigaHertz or 11.70 to 12.75 GigaHertz, switchable)

Mixer

Probe

Low - Noise Amplifier (LNA)

Bandpass Filter

Block IF Amplifier

Block IF Outputs: 950 to 1950 MHz; 1100 to 2150 MHz

Local Oscillator #1 converts the 10.70 to 11.7 GHz Band to Block IF of 950 to 1950 MHz

#1

Local Oscillator (9.75 GHz)

Local Oscillator (10.6 GHz)

#2

Local Oscillator #2 converts the 11.7 to 12.75 GHz Band to Block IF of 1100 to 2150 MHz

Switchable 13/17 volts DC supplied by the receiver up the coaxial cable

Fig. 11-19. A "universal" Ku-band LNB for the European market.

tended C-band frequencies below 3.7 GHz. To do this, these products must generate an extended IF output range from 950 to 1750 MHz. Again, the IF input range of your IRD must match the IF output range of an extended C-band LNB before it can tune to the satellite channels below 3.7 GHz.

Keep in mind that the transponder numbering and the amount of frequency spacing between transponders are not consistent among the various above-mentioned satellite systems. Satellite TV receivers and IRDs which incorporate programmable tuning will give your system added flexibility when pulling in international signals. Refer to the individual satellite transponder charts that appear in Appendix B for further details.

Video Bandwidth. INTELSAT and PANAMSAT sometimes transmit TV signals using only half of a transponder's available bandwidth. This is typically the case when INTELSAT transmits short duration news feeds on one of its global beam transponders. When receiving "half-transponder" signals, the receiver bandwidth must be narrowed from its normal wide setting to less than 20 MHz (INTELSAT) or 27 MHz (PAS) to obtain optimum performance.

Even when a satellite signal occupies the entire transponder, your system may not receive a sufficient signal level to produce a good picture. This is often the case when the antenna is too small for receiving a particular satellite TV service. Many of today's analog receivers and IRDs come with adjustable bandwidth filters built into them. These filters can be adjusted on a channel-by-channel basis to provide optimum reception of each available satellite TV service. Once you have determined the best filter setting, you can store that setting in the receiver's memory circuit

for automatic recall any time that you select that channel for viewing. If you have a satellite receiver that does not have a built-in bandpass filter, you can purchase an outboard filter and connect it to the second IF loop-through ports (usually labeled "IF IN" and "IF OUT") on the receiver's back panel. The frequency of the filter you buy also must match the IF loop-through frequency scheme of the receiver. This information is provided either on the receiver's back panel or in the manual.

While a bandpass filter can be used to remove impulse noise or "sparklies" from your reception of the weaker international satellite TV channels, you will need to readjust the filter setting each time you change channels to get the best possible system performance.

There are limits to the effectiveness of using filters to improve satellite reception. Exact reproduction of video color and detail requires that the satellite receiver process an incoming signal's total transmitted bandwidth. By narrowing the filter's bandwidth, you remove the portion of the signal that contains the greatest amount of noise. The image will be improved, but some color fidelity and picture detail will be lost.

INTERNATIONAL VIDEO STANDARDS

The international community uses several different standards to transmit the picture and color components of video. On a single satellite alone, satellite TV viewers may encounter all three of the basic color encoding systems being used around the world today: NTSC, PAL, and SECAM. (For a refresher course on world video standards, refer back to Chapter 7.) Reception of all three video standards by a satellite TV system will require the use of a multi-standard monitor or a special device that transcodes the TV transmissions from their original standard to one that can be displayed on your regular TV set.

Multi-standard monitors and receivers are available so that you can view overseas satellite TV signals in full color. Some models will automatically detect and display the correct standard. Panasonic manufactures a special VCR that will convert PAL, SECAM, and NTSC signals so that you can record them for later viewing or view them live on any TV set in your home. We would recommend one of these for any first-class international Earth station installation.

You also can receive a black and white picture from any standard on most TV sets. When receiving 625-line video on an NTSC monitor, first readjust the set's vertical hold control to stop the picture from "rolling." Then readjust the vertical linearity control to reduce the picture's height so that the 625-line picture will fit on the screen. The linearity control may be on the back of your set. If this adjustment requires the use of a screwdriver, use an insulated tool to prevent the possibility of shock. For someone receiving NTSC with a 625-line PAL or SECAM TV set, the vertical linearity would be adjusted so that the video fills the entire screen.

Switchable Video De-emphasis. Analog receivers and IRDs have a de-emphasis filter that follows the demodulator stage; slightly different component values are necessary for 525-line video than for 625-line video. If this circuit is not modified, there will be a barely perceptible degradation of the video, more noticeable on marginal signals. Use of the incorrect de-emphasis network also can pose problems when recording satellite TV programs with your VCR or interfere with the re-modulation schemes used in SMATV or CATV RF signal distribution systems. Some international receiver manufacturers provide switchable de-emphasis circuits within their products.

INTERNATIONAL DIGITAL DTH SIGNALS

There are numerous digital DTH signals on the international satellites that are free to air (FTA) services that require no subscription fee or conditional access smart card. To receive these signals you will need to purchase an MPEG-2 DVB-compatible IRD that will allow you to set the unit to the correct transmission parameters for receiving each digital FTA service or bouquet. Manufacturers such as Hyundai, Nokia, and Scientific Atlanta make digital IRDs expressly for tuning in these signals.

The most important settings for any digital IRD are the transponder center frequency and polarization, as well as the symbol rate and forward error correction (FEC) settings for each available digital programming bouquet. The manual controls for adjusting these settings usually are found under the "Installation" or "IRD Set-Up" menu and are displayed as an on-screen graphic. Depending on the digital IRD, the frequency setting may either call for the LNB's equivalent IF Frequency for the service (usually between 950 to 1450 MHz) or the LNB's local oscillator frequency AND the actual satellite frequency (for example: a C-band LNB local oscillator frequency of 5.150 GHz minus the satellite frequency of 4.000 GHz produces an IF frequency of 1150 MHz).

The symbol rate and FEC settings often vary from one digital programming bouquet to the next. On the Hispasat satellite constellation, for example, the digital bouquet on transponder 15 (12.015 GHz, vertical polarization) requires a symbol rate of 27.5

Fig. 11-20. Representative LNB manual adjustment settings for a digital DTH IRD.

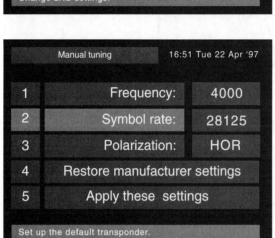

Fig. 11-21. Representative IRD manual adjustment settings for satellite frequency and symbol rate.

All geostationary satellites are launched with enough stationkeeping fuel to counteract the effects of these natural forces for a period of several years, and in the case of the very newest satellites, for fifteen years or longer. It is the precise expenditure of fuel at periodic intervals over the lifetime of each spacecraft that permits each satellite to remain continuously on station at its assigned geostationary orbital position.

The controlling engineers for older INTELSAT V satellites, as well as for all of the Russian Gorizont satellites, typically cease all north/south stationkeeping maneuvers once the satellite has expended most of its remaining fuel. This is done because 90 percent of all fuel expenditures are for making orbital corrections in the north/south direction, while only 10 percent of the fuel is used to make corrections in the east/west direction. A satellite that only has six months of fuel left for normal geostationary operations can have its lifetime extended by several years through the cessation of all north/south stationkeeping maneuvers. The natural forces described above, however, will begin to exert their influence and cause the spacecraft to start drifting away from the Earth's equator in the north/south direction. This is called an "inclined orbit".

Msym/sec and an FEC rate of 3/4. Other services invariably will require their own symbol and FEC rates. Nokia's Mediamaster series will automatically detect the symbol rate and FEC setting in use once you are tuned to the correct transponder frequency and polarization. This particular unit, however, only operates off of 220 volts AC.

TRACKING INCLINED ORBIT SATELLITES

The sun, the moon, and the Earth's gravitational field all affect the movement of satellites orbiting the Earth. These natural effects impel telecommunications satellites to wander from their assigned orbital positions over the Earth's equator. The amount of movement can be predicted, with the direction of movement primarily in the north/south direction over 24-hour intervals.

During the first few months of inclined orbit operations, the amount of drift in the north/south direction is minimal and is transparent to all but the largest receiving antennas on Earth. The amount of drift, however, will continue to increase over time, and the satellite's signal will eventually wander outside of the receiving antenna's main beam for portions of each 24-hour orbital inclination period. The orbital inclination increases at a rate of approximately ± 0.85 degrees per year, and as time goes on, the length of the diurnal (twice each 24-hour period) outages at all receiving stations also will increase.

The only way for a satellite TV system to maintain continuous contact with an inclined orbit satellite is to add a second motor drive to permit the operator to make periodic adjustments to the antenna's declination setting. An adjustment of the antenna's elevation angle alone will not move the antenna along the satellite's north/south path unless the satellite receiving system is located at a longitude that is less than ±10 degrees from the longitudinal location of an inclined orbit satellite.

The least expensive way to track the inclined orbit satellites is to add a second actuator, which replaces the antenna's declination bracket, and an actuator controller. The operator can then manually make adjustments periodically to keep the inclined orbit satellite in view throughout each 24-hour period. This solution is not practical for home satellite TV applications where multiple satellites, including the inclined orbit satellite, will regularly be accessed by the receiving system. Any movement of the polar mount antenna outside of the normal operational boundaries will affect the antenna's ability to properly track geostationary satellites.

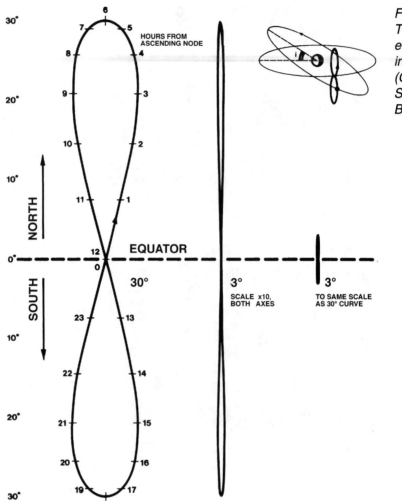

Fig. 11-22. The parameters of an inclined orbit. (Courtesy Stephen J. Birkill.)

Automated receivers and control systems are available that receive data from the satellite receiver's AGC (Automatic Gain Control) in order to obtain information on the relative signal strength of the inclined orbit satellite. If the signal level falls below an established threshold level, the system's microprocessor initiates a search routine that commands the movement of the antenna's tracking system until optimum reception has been restored. These systems also have memory circuits that will automatically return to the correct antenna declination setting for normal satellite reception.

EPILOG:
2001—A SATELLITE ODYSSEY

January 1, 2001—Chiang Mai, Thailand. My last night in the twentieth century started in the midst of a rowdy party in Siem Reap, Cambodia, and ended at the nearby Angkor Wat historical park where I joined thousands of other celebrants to watch the "official" dawn of the new millennium (at least according to the Gregorian calendar). I have it all on digital video tape too, from the chanting Buddhist monks in their flowing saffron robes, to the dancing hill tribe people and awestruck tourists treated to a dazzling sunrise that illuminated the spectacular bas reliefs in Angkor Wat's eastern gallery.

Now that I'm back home, I'll need to transfer today's video onto my multimedia computer system. Like most people who work at home these days, I have a "living office" instead of separate living and office rooms. After all, why duplicate video, stereo audio, and Internet delivery systems in both the office and the living room when a single streamlined system can do the job?

In the 21st century, the distinction between satellite TV receivers and personal computers has all but disappeared. The ongoing global switch from telephone modem to satellite dish for data downloads off the worldwide web is finally making the Internet a viable home entertainment medium. The expansion slots in my new Pentium IV computer are filled with MPEG video, satellite tuner, and smart card reader boards for receiving more than 600 digital DTH TV channels from the various satellites that are available in this part of the world. I

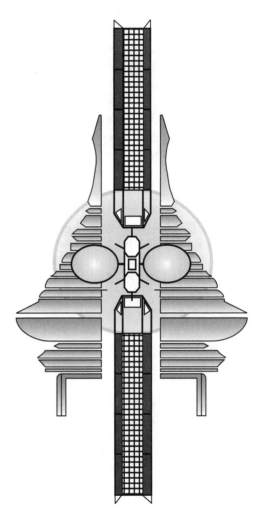

can also download Web sites at dizzying speeds and even access pay-per-view video programs directly from my favorite satellite-based Internet server. Best of all, I can view everything on a new high-definition, digital TV monitor which serves as the heart of my integrated work/play environment.

Much of this was made possible by the global telecom deregulation that occurred at the end of the twentieth century. Several of the newly-privatized national telecommunication companies, as well as their multinational competitors, now offer high-speed satellite access to the Internet with data downloads at dirt cheap prices and dizzying speeds that are thousands of times faster than what the average telephone modem link used to deliver. I still shudder when I recall the days when my typical Internet download made a Bangkok traffic jam look like the Indianapolis 500!

In light of Asia's limited ground-based infrastructure, it was inevitable that certain countries with protective media access policies would eventually be forced to allow individuals to connect their computers to satellite dishes. Each nation's new generation of cybercops continues to control the satellites, of course, so that no unwanted visitors get through the front door.

DIGITAL CONVERGENCE

In case you have been marooned on a desert island for the past year, let me be the first to tell you that on January 15, 2000, a group of university students in Seoul, Korea, founded the "Virtual Channel." An endowment from a couple of the Ka-band satellite operators made it possible for "Virtual [C]" to establish an Internet server onto which anyone can now upload their own digital TV programs. If you have a digital video camcorder and an Internet/MPEG capable multimedia computer system, you too can become a satellite TV program producer!

"Asia's Weirdest Home Videos" was the first Virtual [C] program to attain widespread popularity. More recently, "Eyewitless News" has been in vogue. Amateur videographers now carry their pocket camcorders virtually everywhere in the hope that they will encounter something zany that is also newsworthy so they can get their footage aired on both programs. Several cable TV operators even carry Virtual [C] programming these days. They use satellite dishes to download the Virtual [C] program fare off the Internet and then play out selections according to their own scheduling needs.

If the new Ka-band satellite operators have their way, the distinction between a digital DTH program bouquet and a satellite-based Internet download may soon be history. Why just last month, a new Ka-band satellite was launched to an orbital assignment over New Guinea. The hot news is that for a limited promotional period, the satellite operator is having a fire sale for satellite uplink time. The new Ka-band personal Earth station (PES) that I bought over the Christmas holiday transmits as well as receives satellite signals. It may have been a bit dear at US $3,000, but then again I paid a lot more than that for my first satellite TV system back in 1981.

The new Ka-band satellite on my side of the pond is just one of the many new Ka-band satellites which are jockeying for a share of the new personal "bandwidth on demand" market. Individuals only have to pay for the satellite capacity and airtime that they actually use. The new Ka-band satellite's footprint produces a network of cellular beams, each no more than 400 miles in diameter. One of these "pencil beams" covers my home province quite nicely. To uplink my video, I merely have to find an open transponder frequency and command the satellite to route my program contribution to the Virtual [C] server via the satellite's Seoul, Korea, downlink spot beam. At the push of a button, presto! My new video is on its way to my friends at Virtual [C]. Now all we need to do is make the popcorn, kick back and enjoy.

A MULTIMEDIA REVOLUTION

If the speculations outlined above seem a bit farfetched, they shouldn't. The technologies and trends needed to make this crystal ball vision of the future a reality are already present and accounted for.

Of course, not every technological wonder is an instantaneous hit among consumers. Witness the long time lag between the initial development of high definition television in the mid-1980s and its formal adoption by the FCC in 1996 and the International Telecommunication Union in 1997. The Internet's ever-increasing demand for bandwidth, however, coupled with an inadequate landline infrastructure in many parts of the world, appear to make the marriage of Internet and satellite technologies as close to a sure thing as I can imagine. In Atlanta, Singapore, Paris, or even a sprawling metropolis like Bangkok, fiber-optic cable is one answer to the Internet's voracious appetite for bandwidth. For those of us who live outside a major metropolitan area, however, satellites are the only feasible solution on the horizon.

In today's operating environment, it is as if every local Internet service provider (ISP) is trying to suck up the entire Mississippi River through a soda straw. Satellites are the perfect choice for the implementation of asymmetrical communications networks where the receiving site uploads information requests at low data rates using a telephone line and the transmit site downloads the requested information to a PC at a very high data rate.

Satellite-based Internet servers have been operating in the USA and Europe for a while already. In September of 1997, Zak-Net inaugurated a regional Internet service for the Asia/Pacific region using the AsiaSat 2 satellite. DirecTV Japan followed suit in December of 1997 when it launched a high-speed DirecPC Internet service for Japan. During the early months of 1998, satellite operators the world over sent out a flurry of announcements concerning the launch of a variety of direct-to-home Internet services. The digital DTH bouquets of the new millennium will inevitably be high-speed pipelines that deliver TV AND Internet services to a single dish and "smart" high-definition TV.

Today's pay-TV programmers use automated video servers to format their program lineup at the satellite uplink. These servers consist of bar-coded tape libraries and automated cart machines that insert the tapes in the correct order and at the proper time. In the 21st century, the Internet will also be a video server, but one which the viewer rather than the programmer controls.

TURN ON, PLUG IN AND PLAY OUT

The Internet service providers of the 21st century will also be offering integrated video/data PC cards that plug into the expansion ports of any personal computer system. The installation will be similar in all respects to that of a regular satellite TV receiving system, except that instead of connecting the coaxial cable from the outdoor dish and LNB to a stand-alone receiver, the cable will connect to the back of the computer terminal.

Several new satellite TV products are already targeting the PC "plug and play" operating environment. Germany's Galaxis now offers a complete satellite TV receiver on a PC card. Called the Sat-Surfer PCI, the new product displays PAL or SECAM satellite TV signals in an enhanced resolution, 800 x 600 pixel. Sat-Surfer can also display Teletext from satellite TV sources or even capture individual frames of video. Meanwhile, Hitachi and Pace Microsystems have plans to jointly develop an MPEG-2, DVB-compliant PC card which will allow computer operators to download video, audio and data from a wide variety of sources.

I WANT MY DTV

In the summer of 1997, the ITU formally defined a new universal digital TV standard which combines features from separate digital HDTV standards that America's Advanced Television Standards Committee (ATSC) and Europe's DVB Group have already adopted. The result is a single compatible system that will be implemented by TV set manufacturers worldwide to produce wide-screen TV pictures with a resolution approaching the clarity of 35mm film. The new standard also will offer sixteen sound channels for multilingual broadcasting and support a variety of picture formats including a wide-screen display comprised of 1080 x 1920 pixels.

The new all-digital TV sets are already appearing in the marketplace because digital terrestrial TV is just now beginning in Europe and the USA. Leading TV set manufacturers already have agreed on a common interface that will allow consumers to connect their new digital TV sets to terrestrial, cable and satellite signals. The new digital TV sets also will support a wide range of Conditional Access (CA) systems and software applications. Best of all, there will be no proprietary designs to prevent the new digital TV sets from interacting with any of the available digital program streams.

With its high-resolution video monitor, CD player, and stereo sound system, today's multimedia computer system has become a state-of-the-art home entertainment system. Once the digital TV sets begin to make their appearance, individuals working at home can do so in an integrated work/play environment which I earlier called the "living office."

SPACECRAFT ADVANCEMENTS

In the here and now, one important economic limitation to downloading TV programs off the Internet is the high cost of satellite capacity. At today's prices, a C-band satellite transponder can be leased for about $ 1.5 million per year. This translates into a raw hourly rate of about US $170.

A single wideband (54 MHz or more) satellite transponder could theoretically carry approximately thirty simultaneous movie transmissions at 1.28 Megabits/sec. which translates into a raw cost of $ 8.50 per 90-minute download. With royalty payments, overhead and profit factored in, the cost of delivering pay TV movies on demand over an Internet/Satellite link is quite high in comparison to other delivery options. But that's about to change.

New spacecraft construction technologies are helping to lower the cost of satellite capacity. Given the high reliability of today's electronic circuitry, a geostationary satellite's life in orbit is predominantly a function of the amount of on-board stationkeeping fuel that it takes into space. Until now, every communication satellite has had to carry heavy tanks filled with the hydrazine gas that is used for spacecraft stationkeeping while in orbit. Satellite propulsion systems using new xenon ion technology, however, use the impulses generated by pairs of thrusters that eject electrically-charged particles at a speed of 30 kilometers per second, or nearly ten times the velocity of conventional hydrazine thrusters. Satellite designers can therefore reduce fuel weight by up to 90 percent, which gives a satellite operator several attractive options: the operator can launch a lighter spacecraft at a lower cost; install a more complex, heavier communications payload to reduce the cost per transponder; extend the spacecraft mission lifetime by several years; or any combination thereof.

A $300 million investment typically was required in the mid-1990s to build, insure, launch and operate a sixteen to twenty-four transponder satellite with a ten-year mission lifetime. By 2001, the same $300 million investment will be able to produce a forty-eight or even a seventy-two transponder satellite which achieves a mission lifetime of sixteen years or more.

THE WORLD ABOVE 18 GIGAHERTZ

Several international satellite operators have already announced their plans to launch a new series of Ka-band satellites operating in the 19.2 to 20.2 gigahertz frequency range. A single geostationary Ka-band satellite will be able to simultaneous reuse the available Ka-band frequency spectrum dozens of times by dividing the Earth into a honeycomb of highly focused spot beams, each no more than 400 miles in diameter. An on-board processor will allow users to automatically route their transmissions between any two spot beams or retransmit within the same beam. A single Ka-band satellite will be able to form a virtual Internet in the sky. One such system now under development will be able to relay up to 11,520 duplex circuits operating at a data rate of 384 kilobit/sec. A single Ka-band satellite will therefore be able to support hundreds of thousands of individual sub-scribers who only will need to have access on an occasional-use basis.

What's more, consumers will be able to directly uplink as well as downlink Ka-band satellite signals. Because super-high Ka-band satellite signals have such tiny wavelengths, the beam width produced by each personal uplink antenna will be so narrow that the interference problems which plague satellites operating in the lower satellite frequency bands will be almost nonexistent, even when small dishes are used down on the ground. Putting the uplink under the personal control of each subscriber also helps to cut down on operational costs; the subscriber doesn't have to pay to send the signal to an uplink or to help bear the cost of supporting a turnaround facility.

If you thought that the Ka-band represented the new satellite frontier you'd better hold on to your hat. One satellite operator recently filed an application with the FCC to construct, launch and operate a series of twelve geostationary satellites, with downlink signals using even higher satellite frequencies.

To reach fruition, all the technological wonders described above ultimately will depend on what the global economic and regulatory environments are like by the time the year 2001 rolls around. Fortunately, there are now international telecommunication agreements in place that guarantee the opening of previously isolated markets worldwide within the first few years of the new millennium. By then, as we Americans like to say, it will be a whole new ball game.

Mark Long
On the road in Southeast Asia
April 23, 1998

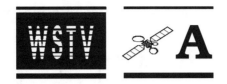

APPENDIX A:
SATELLITE COVERAGE MAPS

The following Ku-band and C-band computational charts and satellite footprint maps can be used to approximate the antenna size that will be required at your location to receive any analog or digital DTH TV service. Keep in mind, however, that the EIRP contour values provided on the maps herein are for satellite services that operate at full transponder saturation—that is, at the maximum power level that the satellite transponder was designed to deliver. In the case of some international satellite transmissions, transponder power may be divided amongst several different communications signals. Moreover, many of these maps were generated prior to the launch of the satellite and therefore should be regarded as projections of what the satellite manufacturer intended to achieve rather than the real-world conditions. Whenever possible, supplement the data provided in the following section with feedback from other satellite users in your area or from the various Internet sources that are available on the worldwide web. To access links to current sources of information for your area, visit my web site at: http://www.mlesat.com.

Each chart assumes nominal performance characteristics for the receiving system antenna, LNF/LNB and IRD. The charts also assume that the satellite or satellites that you intend to view are relatively high in the sky versus near to the horizon. Remember the noise temperature of the receiving antenna is a function of its diameter AND the elevation angle of the dish. The amount of noise entering the antenna system also will depend on the distribution of the antenna's side lobes, the type of feed in use, and other factors.

RAIN OUTAGES

The Ku-band satellite EIRP to antenna diameter conversion chart assumes clear sky conditions. Signal attenuation, weakening due to moderate to heavy rainfall, should be taken into account when designing any Ku-band satellite receiving system. Raindrops can scatter or absorb satellite signals as the spherical wavefront of the transmitted microwave energy radiates Earthward through the lower atmosphere.

Rain attentuation becomes a major consideration when the wavelength of the transmitted microwave signal approaches the size of the raindrop. Rain attenuation significantly affects the satellite frequency bands above 10 GHz. The presence of rain in the sky will also raise the noise temperature of the receiving system because the rain itself has a noise temperature of 290 K versus an average clear sky noise temperature of 17 K. The only effective way to compensate for the attenuating effects of heavy rainfall is to increase the size of the receiving antenna. This will provide a signal margin of several dB to compensate for the signal attenuation that takes place when moderate to heavy rainfall occurs in your area.

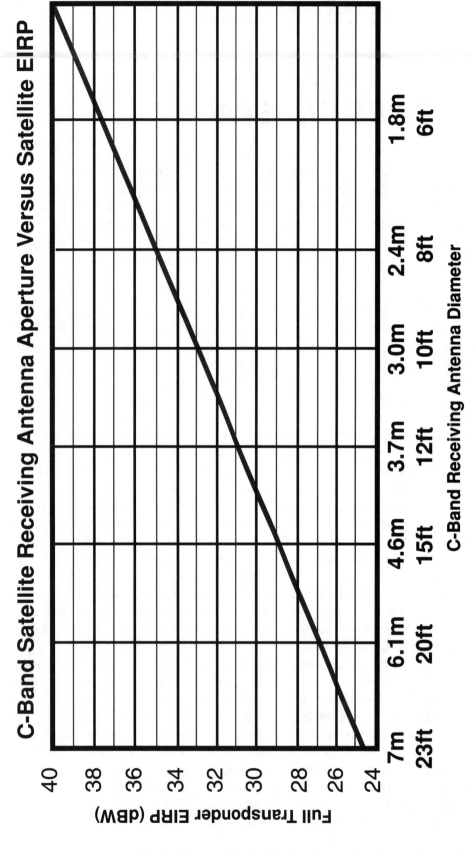

Fig. A-1. C-band antenna size calculation chart.

Fig. A-2. Ku-band antenna size calculation chart.

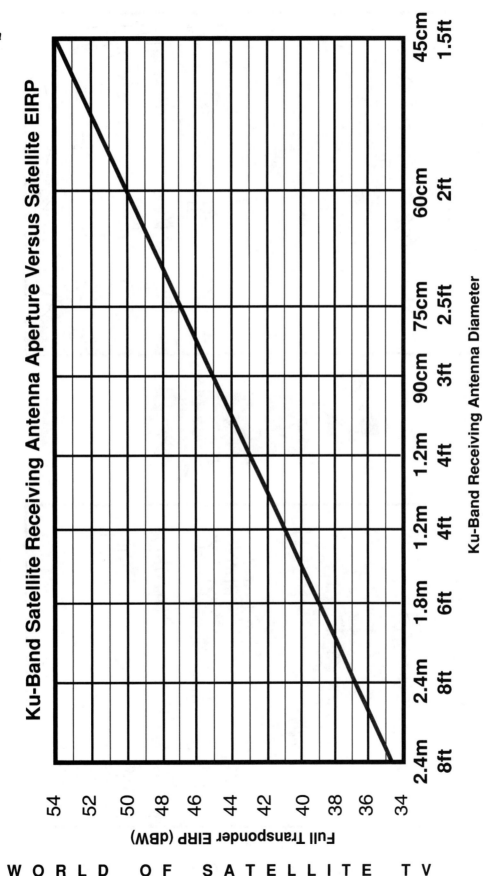

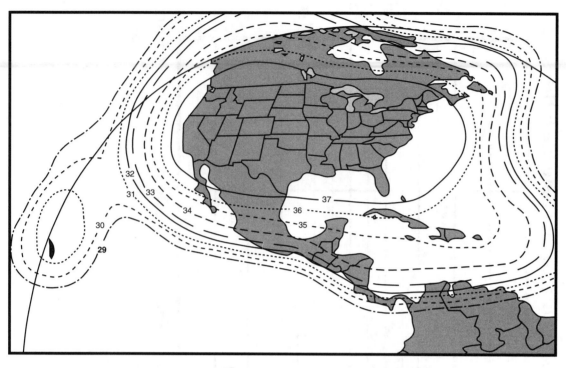

Fig. A-3. GE-2 C-band transmit EIRP (in dBW) from 85° west longitude.

Fig. A-4. GE-2 Ku-band transmit EIRP (in dBW) from 85° west longitude.

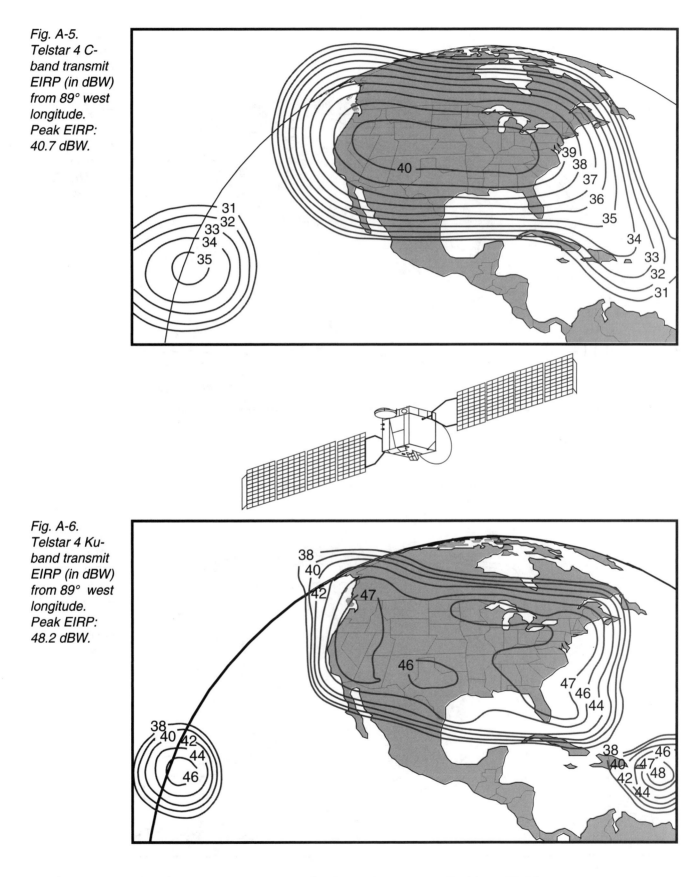

Fig. A-5. Telstar 4 C-band transmit EIRP (in dBW) from 89° west longitude. Peak EIRP: 40.7 dBW.

Fig. A-6. Telstar 4 Ku-band transmit EIRP (in dBW) from 89° west longitude. Peak EIRP: 48.2 dBW.

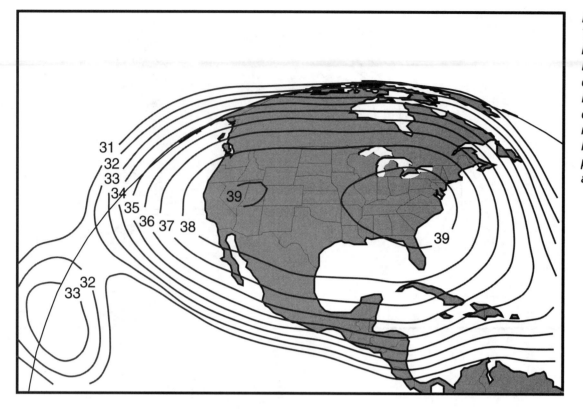

Fig. A-7. Telstar 6 C-band transmit EIRP (in dBW). Peak EIRP: 39.5 dBW. (Preliminary map, location pending FCC approval.)

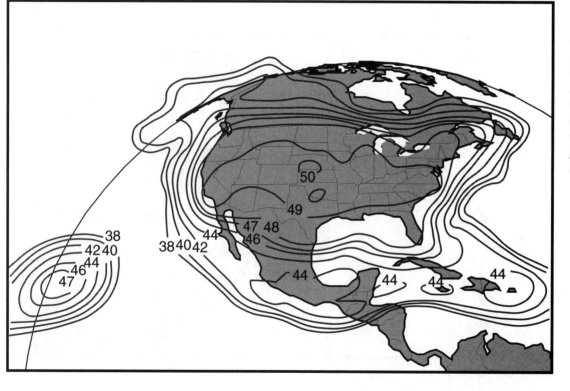

Fig. A-8. Telstar 6 Ku-band transmit EIRP (in dBW). Peak EIRP: 50.5 dBW. (Preliminary map, location pending FCC approval.)

Fig. A-9.
Telstar 5 C-band transmit EIRP (in dBW) from 97° west longitude. Peak EIRP: 39.5 dBW.

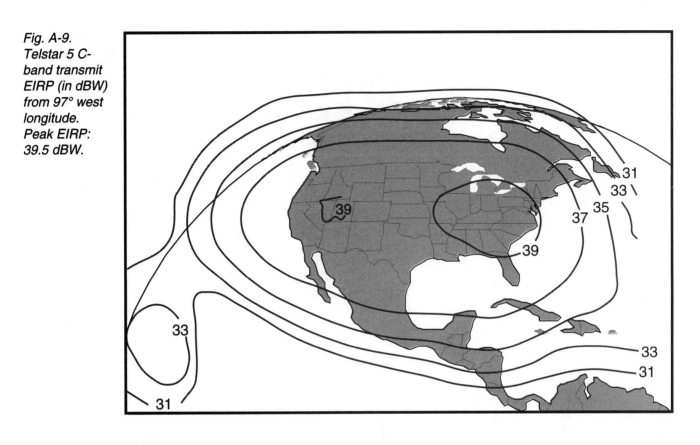

Fig. A-10.
Telstar 5 Ku-band transmit EIRP (in dBW) from 97° west longitude.

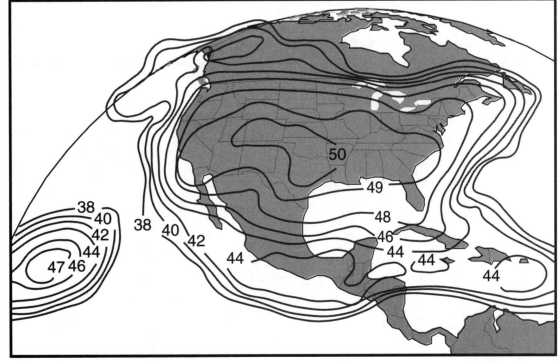

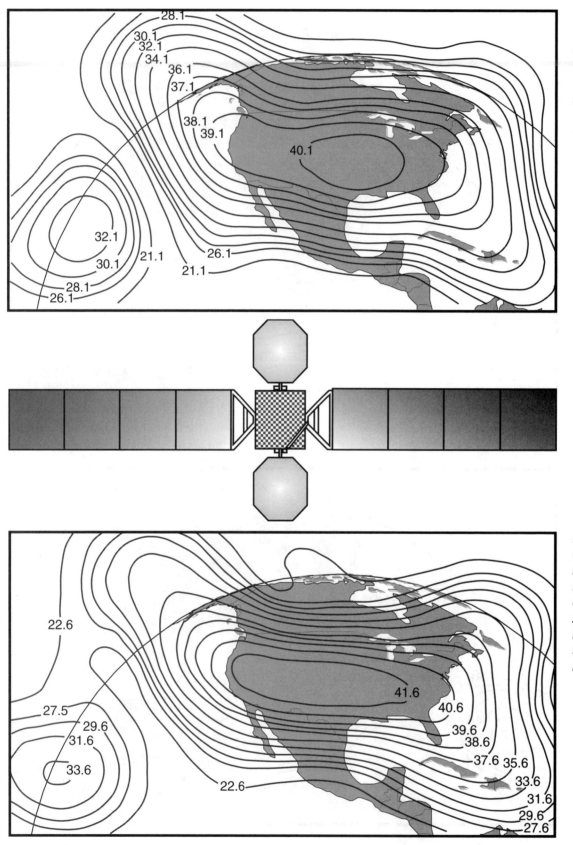

Fig. A-11. Galaxy IV C-band transmit EIRP (in dBW), vertical polarization, from 99° west longitude.

Fig. A-12. Galaxy IV C-band transmit EIRP (in dBW), horizontal polarization, from 99° west longitude.

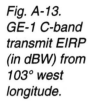

Fig. A-13. GE-1 C-band transmit EIRP (in dBW) from 103° west longitude.

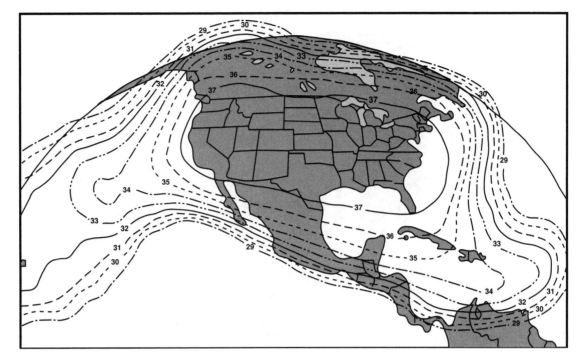

Fig. A-14. GE-1 Ku-band transmit EIRP (in dBW) from 103° west longitude.

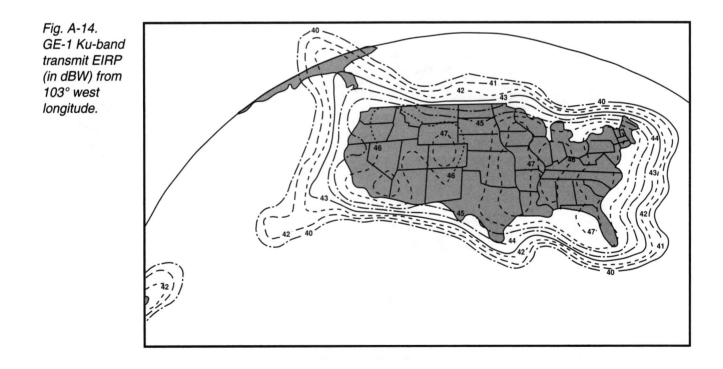

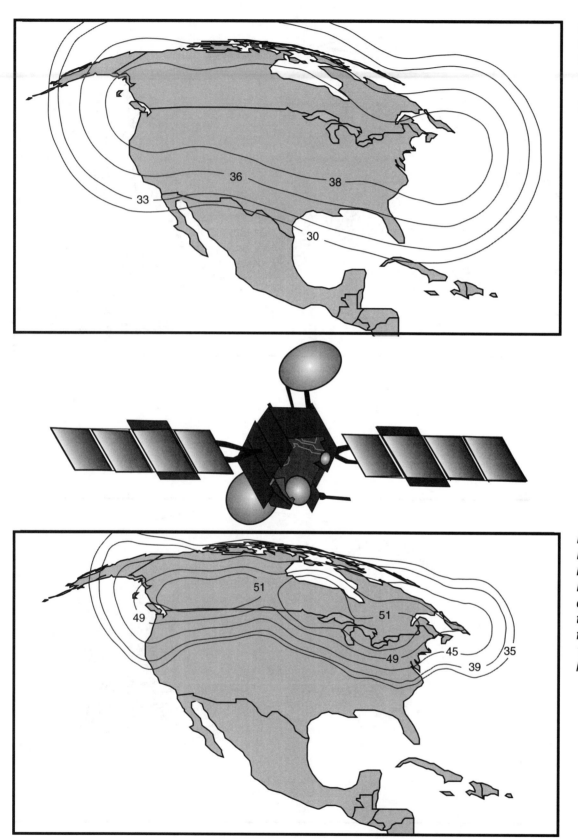

Fig. A-15. Anik E2 C-band transmit EIRP (in dBW) from 107.3° west longtiude.

Fig. A-16. Anik E2 typical Ku-band transmit EIRP (in dBW), National Beam, from 107.3° west longitude.

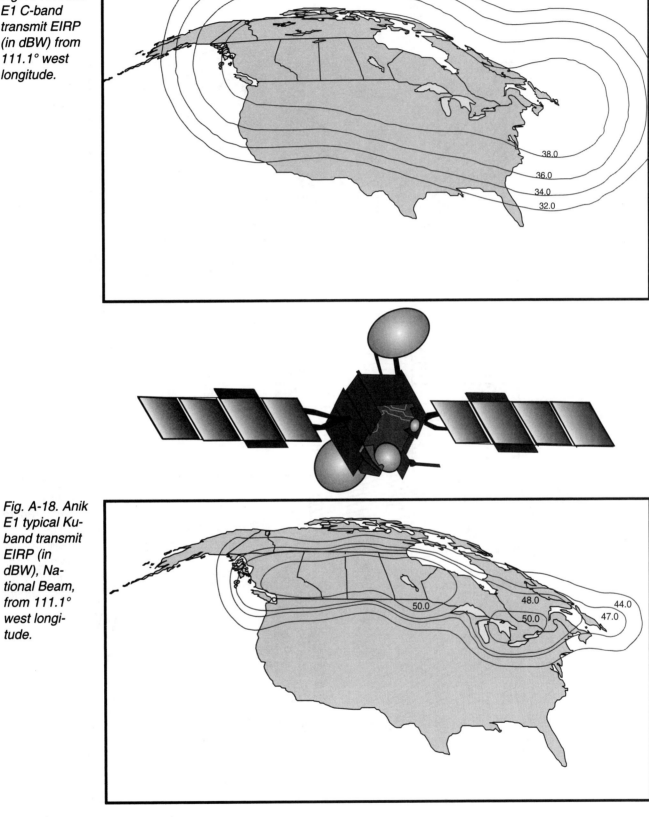

Fig. A-17. Anik E1 C-band transmit EIRP (in dBW) from 111.1° west longitude.

Fig. A-18. Anik E1 typical Ku-band transmit EIRP (in dBW), National Beam, from 111.1° west longitude.

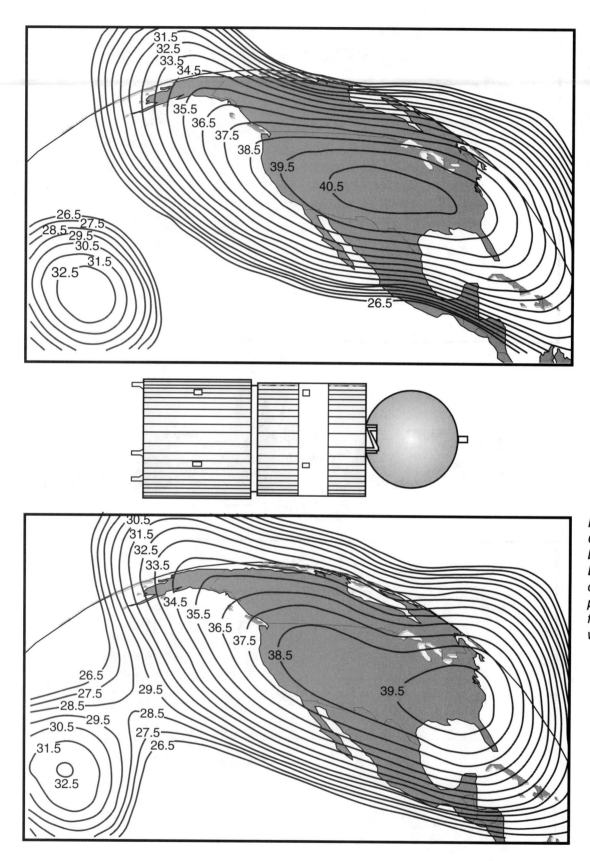

Fig. A-19. Galaxy V C-band transmit EIRP (in dBW), horizontal polarization, from 125° west longitude.

Fig. A-20. Galaxy V C-band transmit EIRP (in dBW), vertical polarization, from 125° west longitude

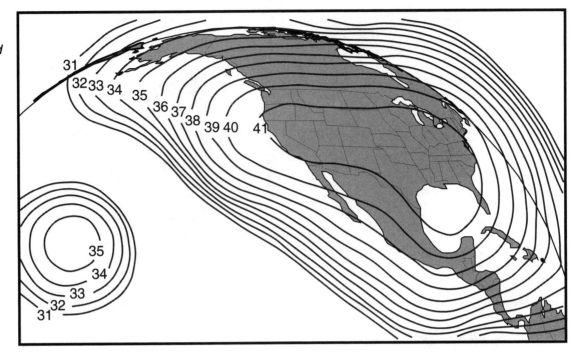

Fig. A-21. Telstar 7 C-band projected transmit EIRP (in dBW), pending FCC approval.

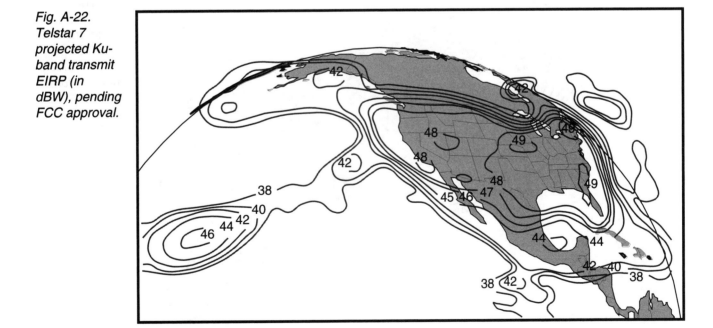

Fig. A-22. Telstar 7 projected Ku-band transmit EIRP (in dBW), pending FCC approval.

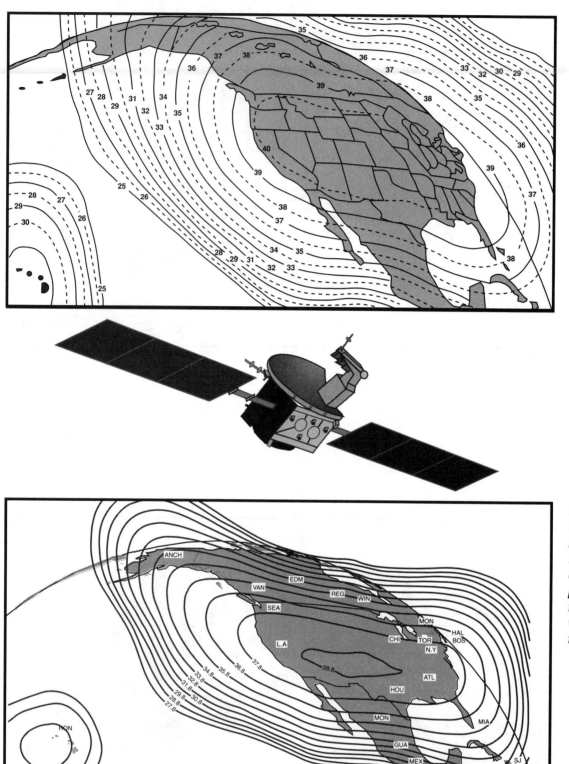

Fig. A-23. Satcom C3 and Satcom C4 C-band transmit EIRP (in dBW) from 131° and 135° west longitude, respectively.

Fig. A-24. Galaxy 1-R C-band transmit EIRP (in dBW), vertical polarization, from 133° west longitude.

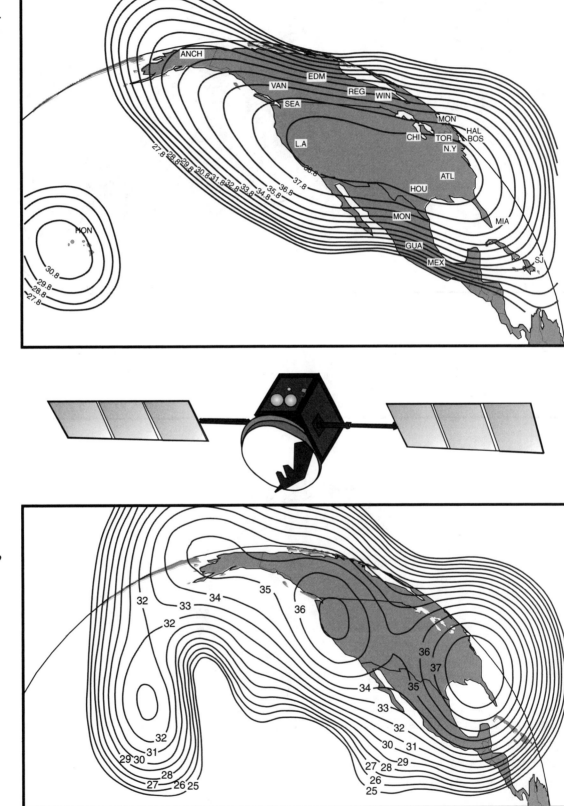

Fig. A-25. Galaxy 1-R C-band transmit EIRP (in dBW), horizontal polarization, from 133° west longitude.

Fig. A-26. Satcom C1 C-band transmit EIRP (in dBW) at 137° west longitude.

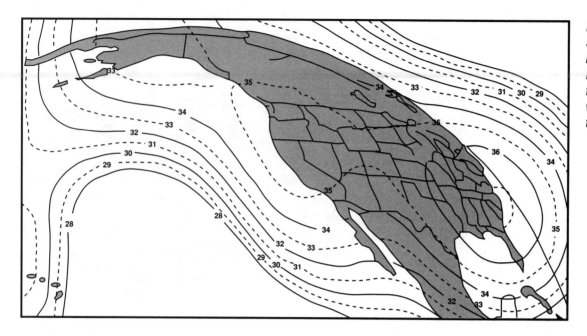

Fig. A-27. Satcom C5 C-band transmit EIRP (in dBW) from 139° west longitude.

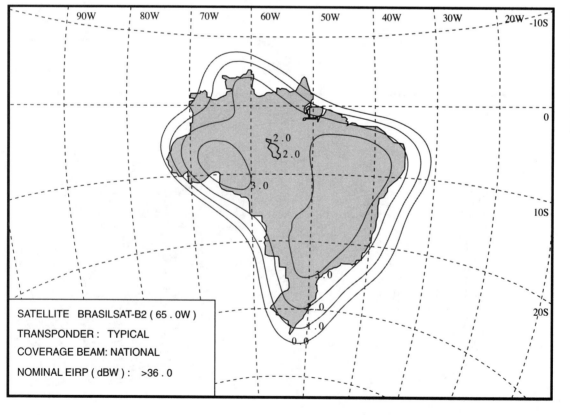

SATELLITE BRASILSAT-B2 (65 . 0W)

TRANSPONDER : TYPICAL

COVERAGE BEAM: NATIONAL

NOMINAL EIRP (dBW) : >36 . 0

Fig. A-28. Brasilsat B2 C-band national beam transmit EIRP (in dBW) from 65° west longitude.

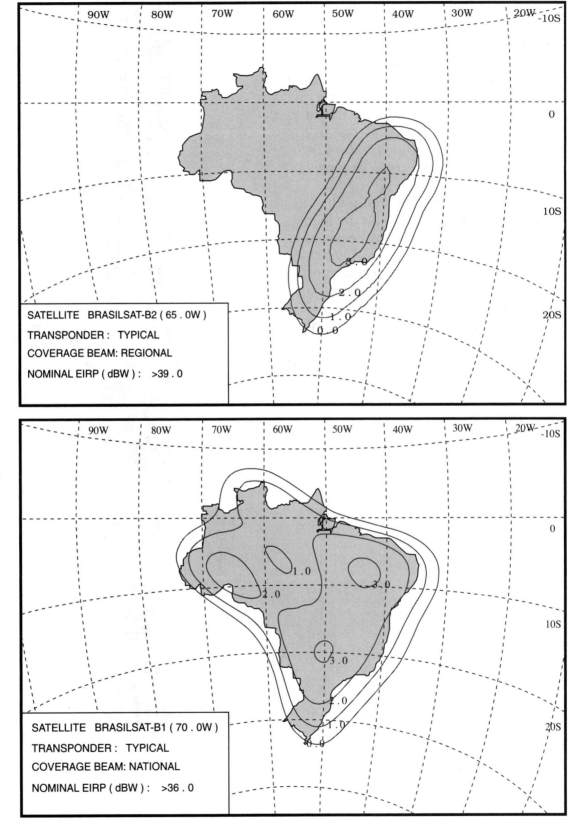

*Fig. A-29.
Brasilsat B2
C-band
regional beam
transmit EIRP
(in dBW) from
65° west
longitude.*

SATELLITE   BRASILSAT-B2 (65 . 0W)

TRANSPONDER :   TYPICAL

COVERAGE BEAM: REGIONAL

NOMINAL EIRP (dBW) :   >39 . 0

*Fig. A-30.
Brasilsat B1
C-band
regional beam
transmit EIRP
(in dBW) from
70° west
longitude.*

SATELLITE   BRASILSAT-B1 (70 . 0W)

TRANSPONDER :   TYPICAL

COVERAGE BEAM: NATIONAL

NOMINAL EIRP (dBW) :   >36 . 0

PAS-5 is the first of a new generation of Hughes HS-601 HP (for High Power) flight models off the production line. The spacecraft is worthy of special mention because it is the first commercial communications satellite to carry the xenon ion propulsion system that was mentioned in the Epilog as one of the major satellite innovations that will revolutionize the industry in the new millennium.

The spacecraft carries multiple beams, a few of which are shown on the following pages. The satellite was designed to deliver DTH services to Latin America as well as provide video and telecommunications services for the Americas and Europe. The spacecraft carries 24 C-band transponders with 50-watt amplifiers and 24 Ku-band transponders, six with 60-watt amplifiers, and eighteen with 110-watt amplifiers.

Fig. A-31. PAS-5 North American Ku-band beam (1) transmit EIRP (in dBW) from 58° west longitude.

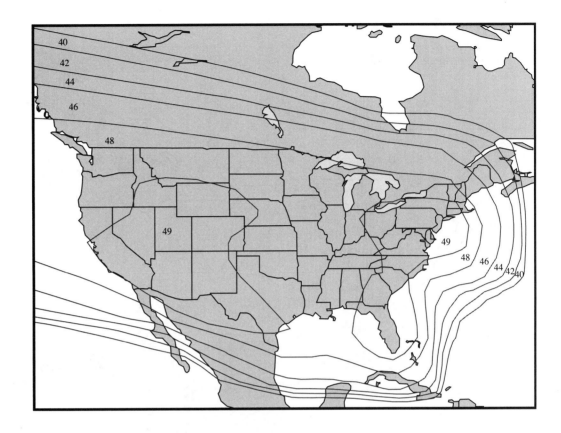

Fig. A-32. PAS-5 North American Ku-band beam (2) transmit EIRP (in dBW) from 58° west longitude.

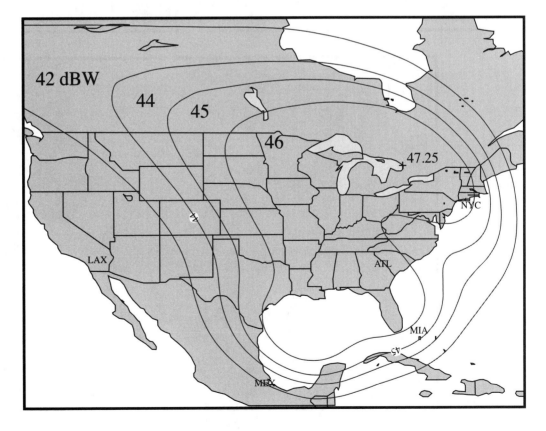

T H E   W O R L D   O F   S A T E L L I T E   T V   213

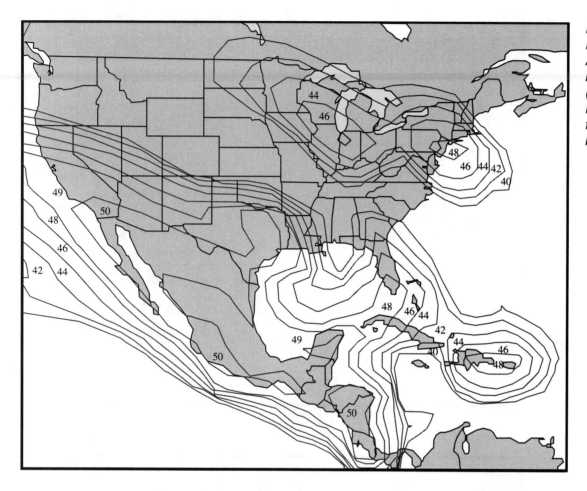

Fig. A-33. PAS-5 North American Ku-band beam (3) transmit EIRP (in dBW) from 58° west longitude.

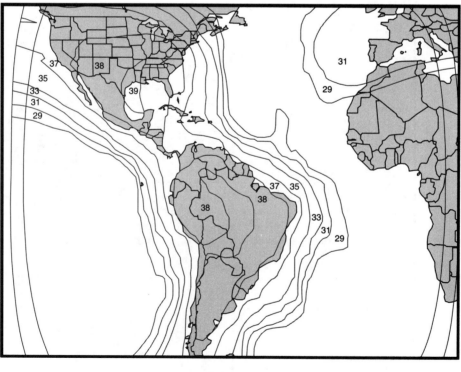

Fig. A-34. PAS-5 C-band beam transmit EIRP (in dBW) from 58° west longitude.

Fig. A-35.
INTELSAT
706 C-band
global beam
coverage
from 53° west
longitude.
Beam edge
EIRP: 26
dBW,
with contour
values of 0,
+1, +2, +3 &
+4.

Fig. A-36.
INTELSAT
706 C-band
hemi beam
coverage from
53° west
longitude.
Beam edge
EIRP: 33
dBW, with
contour values
of 0, +1, +2,
+3 , +4 & +5.

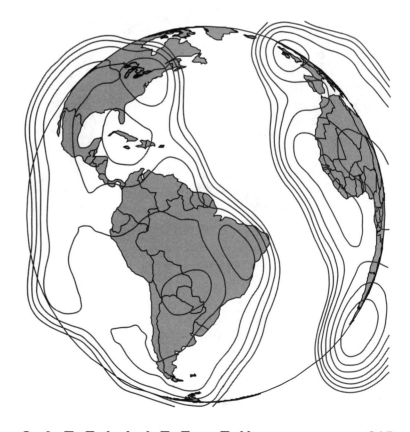

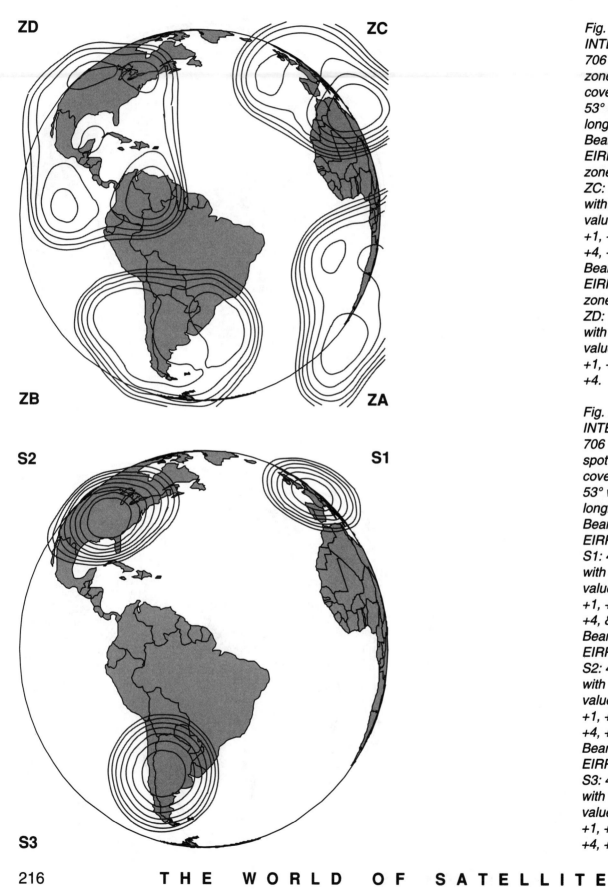

ZD

ZC

ZB

ZA

S2

S1

S3

Fig. A-37. INTELSAT 706 C-band zone beam coverage from 53° west longitude. Beam edge EIRP for zones ZA & ZC: 33 dBW, with contour values of 0, +1, +2, +3, +4, +5 & +6. Beam edge EIRP for zones ZB & ZD: 33.5 dBW, with contour values of 0, +1, +2, +3, & +4.

Fig. A-38. INTELSAT 706 Ku-band spot beam coverage from 53° west longitude. Beam edge EIRP for Spot S1: 43.5 dBW, with contour values of 0, +1, +2, +3, +4, & +5. Beam edge EIRP for Spot S2: 42.5 dBW, with contour values of 0, +1, +2, +3, +4, +5 & +6. Beam edge EIRP for Spot S3: 44 dBW, with contour values of 0, +1, +2, +3, +4, +5 & +6.

Fig. A-39.
INTELSAT
706 C-band
spot beam
coverage from
53° west
longitude.
Beam edge
EIRP: 33
dBW, with
contour values
of 0, +1, +2,
+3 , & +4.

Fig. A-40.
INTELSAT
709 C-band
spot beam
coverage from
50° west
longitude.
Beam edge
EIRP: 33
dBW, with
contour values
of 0, +1, +2,
+3 , & +4.

THE WORLD OF SATELLITE TV

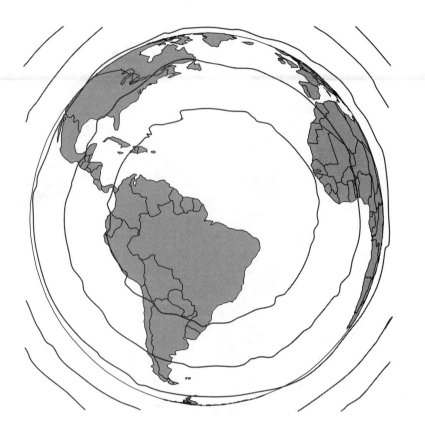

Fig. A-41. INTELSAT 709 C-band global beam coverage from 50° west longitude. Beam edge EIRP: 26 dBW, with contour values of 0, +1, +2, +3 & +4.

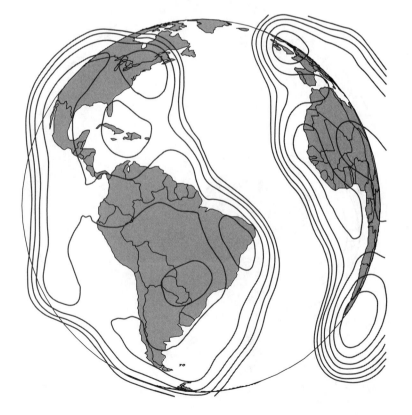

Fig. A-42. INTELSAT 709 C-band hemi beam coverage from 50° west longitude. Beam edge EIRP: 33 dBW, with contour values of 0, +1, +2, +3 , +4 & +5.

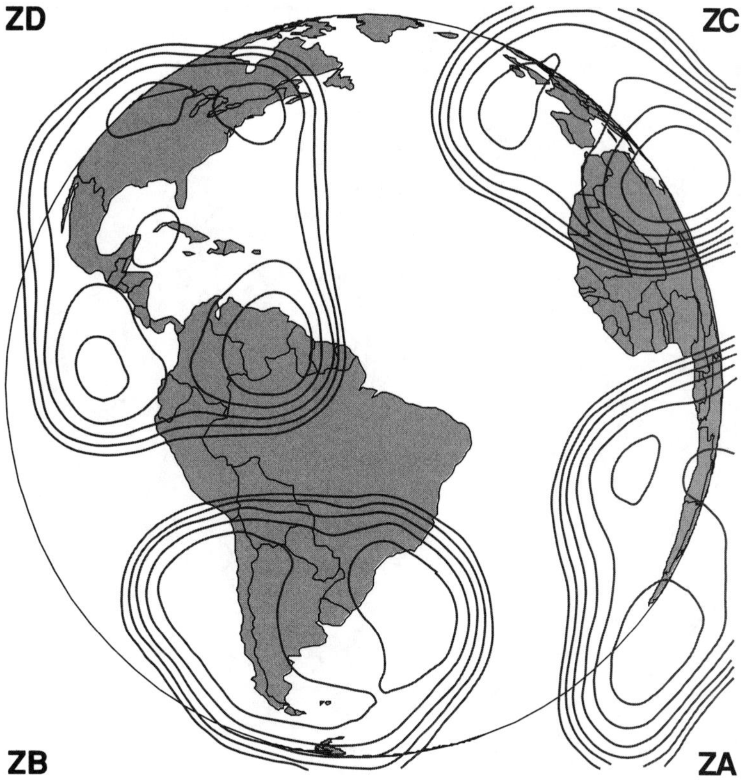

Fig. A-43.
INTELSAT
709 C-band
zone beam
coverage from
50° west
longitude.
Beam edge
EIRP for
zones ZA &
ZC: 33 dBW,
with contour
values of 0,
+1, +2, +3 ,
+4, +5 & +6.
Beam edge
EIRP for
zones ZB &
ZD: 33.5 dBW,
with contour
values of 0,
+1, +2, +3, &
+4.

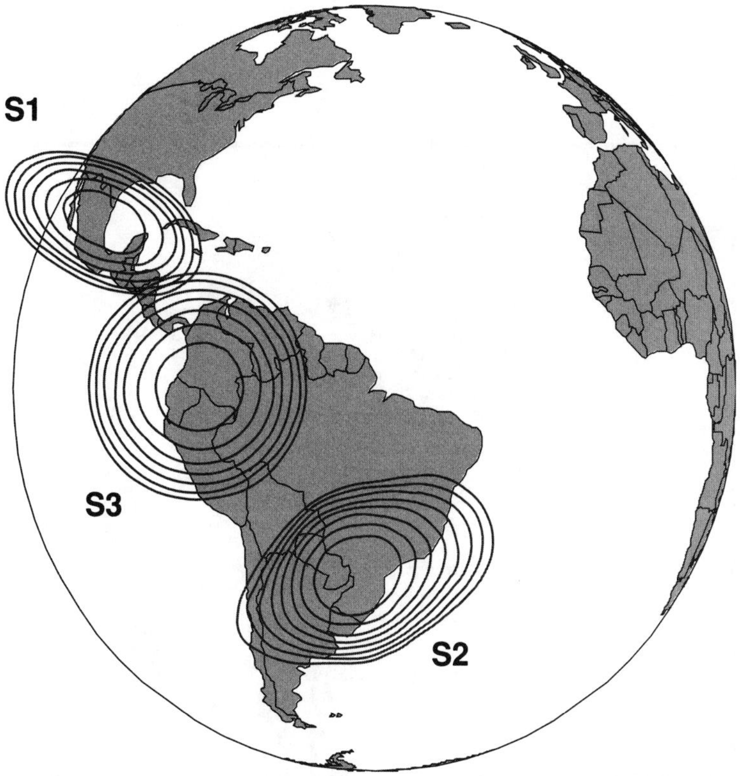

Fig. A-44.
INTELSAT
709 Ku-band
spot beam
coverage from
50° west
longitude.
Beam edge
EIRP for Spot
S1: 43.5 dBW,
with contour
values of 0,
+1, +2, +3 ,
+4, & +5.
Beam edge
EIRP for Spot
S2: 42.5 dBW,
with contour
values of 0,
+1, +2, +3 ,
+4, +5 & +6.
Beam edge
EIRP for Spot
S3: 44 dBW,
with contour
values of 0,
+1, +2, +3 ,
+4, +5 & +6.

T H E   W O R L D   O F   S A T E L L I T E   T V

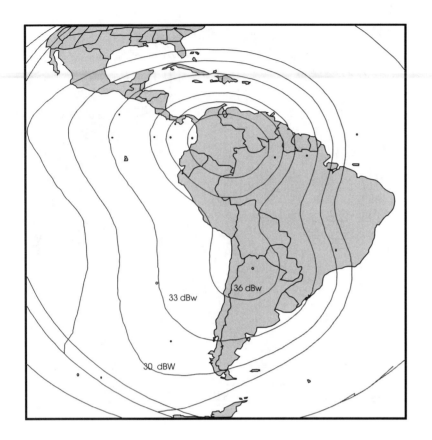

33 dBw

36 dBw

30 dBW

*Fig. A-45.
PAS-1 C-band
Latin Ameri-
can beam
transmit EIRP
(in dBW) from
45° west
longitude.*

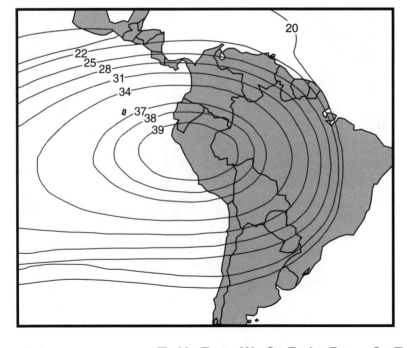

20

22
25
28
31
34
37
38
39

*Fig. A-45.
PAS-1 C-band
central beam
transmit EIRP
(in dBW) from
45° west
longitude.*

Fig. A-47. PAS-1 Ku-band North American beam transmit EIRP (in dBW) from 45° west longitude.

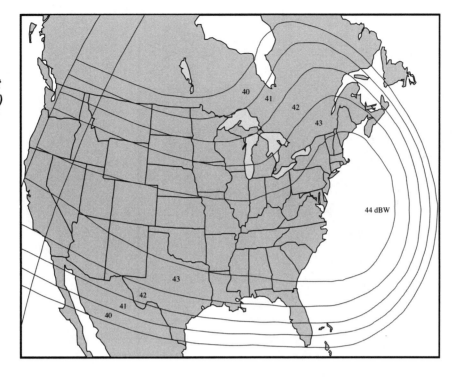

Fig. A-48. PAS-1 C-band northern beam transmit EIRP (in dBW) from 45° west longitude.

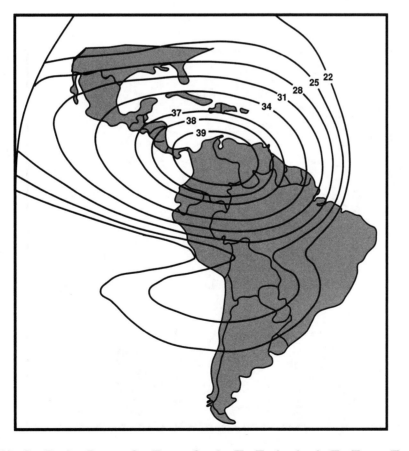

THE WORLD OF SATELLITE TV 221

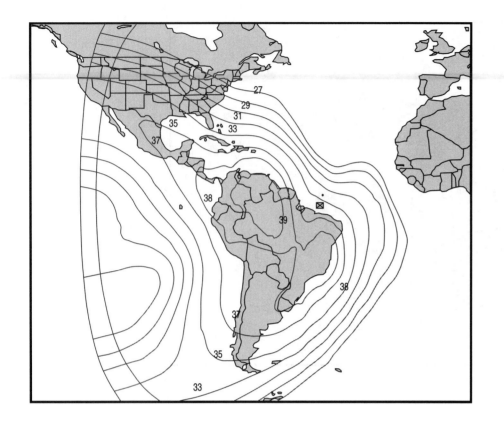

Fig. A-49. PAS-3R C-band Americas (1) transmit EIRP (in dBW) from 43° west longitude.

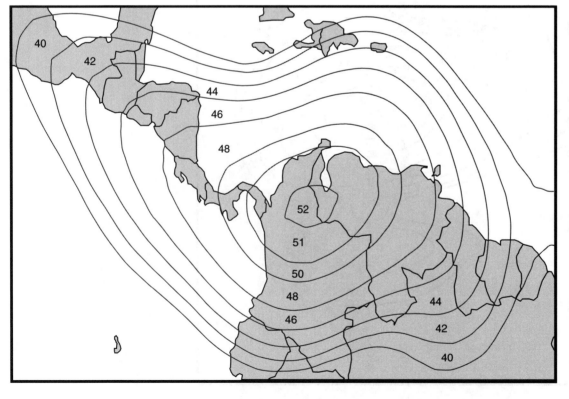

Fig. A-50. PAS-3R Ku-band northern beam transmit EIRP (in dBW) from 43° west longitude.

Fig. A-51. PAS-3R C-band Americas (2) transmit EIRP (in dBW) from 43° west longitude.

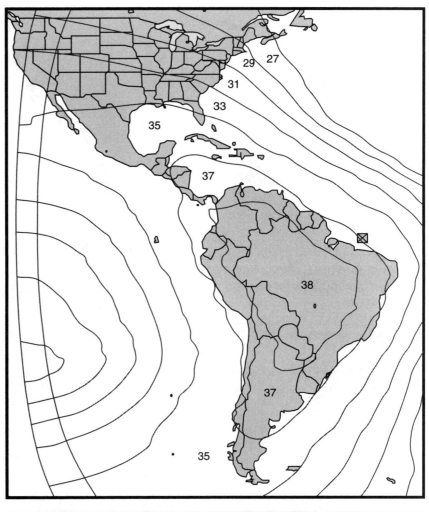

Fig. A-52. PAS-3R Ku-band North America (1) beam transmit EIRP (in dBW) from 43° west longitude.

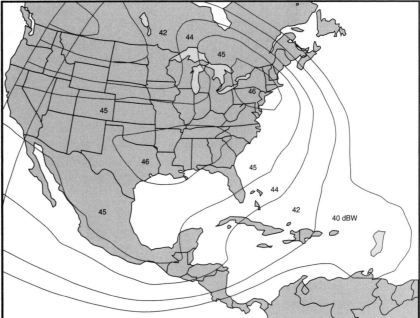

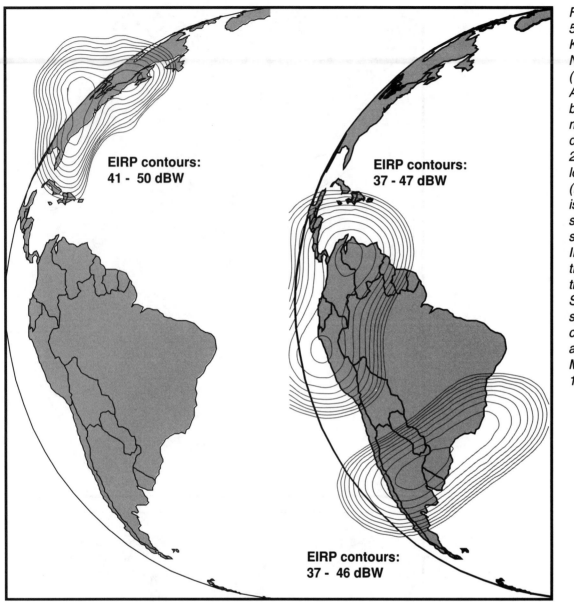

EIRP contours:
41 - 50 dBW

EIRP contours:
37 - 47 dBW

EIRP contours:
37 - 46 dBW

Fig. A-53 & A-54. INTELSAT K Ku-band North America (1) and South America (2) beams transmit EIRP (in dBW) from 21.5° west longitude. (INTELSAT K is one of several satellites that INTELSAT is transferring to the "New Skies" public spin-off company announced in March of 1998.)

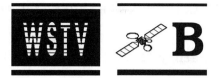

APPENDIX B: SATELLITE TRANSPONDER FREQUENCY PLANS

Knowledge of the correct satellite frequency, polarization and beam coverage area is an important part of searching for any TV or radio services that may be available on any one of the numerous satellites now serving the Americas. Although the C-band frequency plans for most North American domsats conform to a standard plan (see Fig. 6-9 in Chapter 6), several different Ku-band plans are in use within the region. There also are more than two dozen international satellites that can be received from locations in the Americas. Most of these satellites use transponder center frequencies and bandwidths that vary from one type of satellite to the next. Many satellites also have beam switching capabilities so that designated transponders can be connected to any one of several coverage beams.

The following C-band and Ku-band satellite transponder plans represent the vast majority of the satellites now available. Two sets of frequencies are presented on many of these charts. One set of frequencies represents the uplink channels (5.95 to 6.45 GHz for C-band and 14.0 to 14.5 GHz for Ku-band) used to send signals to the satellite, while the other set represents the downlink frequencies to which all satellite Earth stations must tune in order to receive the transmissions. The receiving Earth station also must be set to the correct polarization for the transponder under scrutiny as well as be located within the satellite coverage beam in use.

Because of the high demand for satellite time, international satellite organizations such as INTELSAT will often squeeze two analog television signals—or one TV signal and a number of non-video communications channels—onto a single satellite transponder. Tuning to any given satellite transponder's center frequency may therefore not immediately reveal the presence of TV signals. The center frequency for a satellite transponder should be regarded merely as the starting point for any search.

For example, INTELSAT regularly delivers analog TV signals in a "half-transponder" format on the top C-band transponder (4.175 GHz) of its AOR satellites located at 18.5, 24.5 and 34.5 degrees west longitude. In this case, the center frequencies of the half-transponder TV signals are 4.1665 GHz and 4.1885 GHz. Use the frequency adjustment control provided by the receiver or IRD to tune through the entire bandwidth of the transponder.

When receiving half-transponder analog TV signals, it is important to have the bandwidth of the receiver or IRD narrowed to less than 20 MHz. If the set-top box that you are using does not provide any filter adjustment options, narrowband filters are available that can be connected to the second IF loopthrough connections provided on the back of most analog receivers and IRDs. Narrowing the receive passband removes noise that would otherwise affect your reception of the signal.

U.S. DBS Transponder Frequency Chart
Bandwidth: 27 MHz

Tr. No.	Pol.	Frequency (in GHz)
01	RH	12.224
02	LH	12.239
03	RH	12.253
04	LH	12.268
05	RH	12.282
06	LH	12.297
07	RH	12.311
08	LH	12.326
09	RH	12.341
10	LH	12.355
11	RH	12.370
12	LH	12.384
13	RH	12.399
14	LH	12.414
15	RH	12.428
16	LH	12.443
17	RH	12.457
18	LH	12.472
19	RH	12.486
20	LH	12.501
21	RH	12.516
22	LH	12.530
23	RH	12.545
24	LH	12.559
25	RH	12.574
26	LH	12.588
27	RH	12.603
28	LH	12.618
29	RH	12.632
30	LH	12.647
31	RH	12.661
32	LH	12.676

ANIK E Transponder Frequency Chart
Bandwidth: 27 MHz

Tr. No.	Pol.	Frequency
01	V	11.717
02	V	11.743
03	V	11.778
04	V	11.804
05	V	11.839
06	V	11.865
07	V	11.900
08	V	11.926
09	V	11.961
10	V	11.987
11	V	12.022
12	V	12.048
13	V	12.083
14	V	12.109
15	V	12.144
16	V	12.170
17	H	11.730
18	H	11.756
19	H	11.791
20	H	11.817
21	H	11.852
22	H	11.878
23	H	11.913
24	H	11.939
25	H	11.974
26	H	12.000
27	H	12.035
28	H	12.061
29	H	12.096
30	H	12.122
31	H	12.157
32	H	12.183

Galaxy 4 Transponder Frequency Chart
Bandwidth: H-27 MHz V-54 MHz

Tr. No.	Pol.	Frequency
01	V	11.720
02	H	11.750
03	V	11.750
04	H	11.780
05	V	11.810
06	H	11.810
07	V	11.840
08	H	11.870
09	V	11.870
10	H	11.900
11	V	11.930
12	H	11.930
13	V	11.960
14	H	11.990
15	V	11.990
16	H	12.020
17	V	12.050
18	H	12.050
19	V	12.080
20	H	12.110
21	V	12.110
22	V	12.140
23	H	12.170
24	V	12.170

Fig. B-1 & B-2. North American Ku-band DBS and digital DTH satellite transponder frequency plans.

Galaxy 7 Transponder Frequency Chart
Bandwidth: H-27 MHz V-54 MHz

01	V	11.720
02	H	11.750
03	V	11.750
04	H	11.780
05	V	11.810
06	H	11.810
07	V	11.840
08	H	11.870
09	V	11.870
10	V	11.900
11	V	11.945
12	V	11.930
13	V	11.960
14	H	11.976
15	V	11.991
16	V	12.020
17	H	12.065
18	V	12.050
19	V	12.080
20	H	12.110
21	V	12.110
22	V	12.140
23	H	12.170
24	V	12.170

GE-1, GE-2 & GE-3
Transponder Frequency Chart
Bandwidth: 36 MHz

01	H*	11.720
02	V*	11.740
03	H*	11.760
04	V*	11.780
05	H*	11.800
06	V*	11.820
07	H*	11.840
08	V*	11.860
09	H*	11.880
10	V*	11.900
11	H*	11.920
12	V*	11.940
13	H*	11.960
14	V*	11.980
15	H*	12.000
16	V*	12.020
17	H*	12.040
18	V*	12.060
19	H*	12.080
20	V*	12.100
21	H*	12.120
22	V*	12.140
23	H*	12.160
24	V*	12.180

*polarization reversed on GE-3

SBS 6 Transponder Frequency Chart
Bandwidth: 43 MHz

01	H	11.725.0
02	V	11.749.5
03	H	11.774.0
04	V	11.798.5
05	H	11.823.0
06	V	11.847.5
07	H	11.872.0
08	V	11.896.5
09	H	11.921.0
10	V	11.945.5
11	H	11.970.0
12	V	11.994.5
13	H	12.019.0
14	V	12.043.5
15	H	12.068.0
16	V	12.092.5
17	H	12.117.0
18	V	12.141.5
19	H	12.166.0

Telstar 4 Transponder Frequency Chart
Bandwidth: 54 MHz

01	V	11.730
02	H	11.743
03	V	11.790
04	H	11.803
05	V	11.850
06	H	11.863
07	V	11.910
08	H	11.923
09	V	11.971
10	H	11.984
11	V	12.033
12	H	12.046
13	V	12.095
14	H	12.108
15	V	12.157
16	H	12.170

INTELSAT 801 - 804 SATELLITES
C-Band Transponders

Fig. B-3.
INTELSAT VIII
C-band
satellite
transponder
frequency
plan.

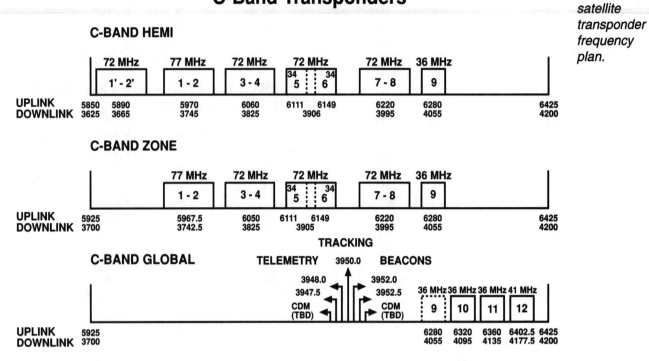

Fig. B-4.
INTELSAT VIII
(801 - 804)
Ku-band
satellite
transponder
frequency
plan.

INTELSAT 801 - 804 SATELLITES

Ku-Band Transponders

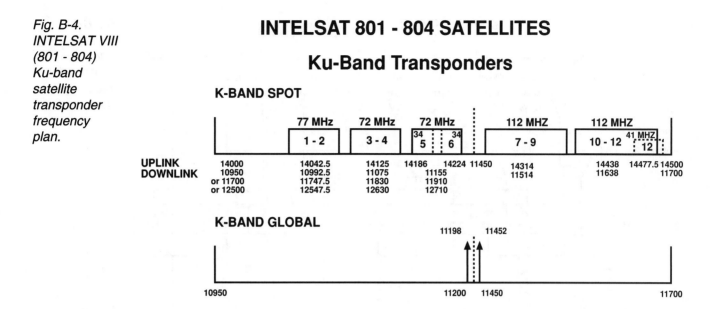

NOTE: *TRANSPONDER 1' - 2' IS ONLY AVAILABLE FOR HEMI BEAM OPERATION

EXPRESS Satellite

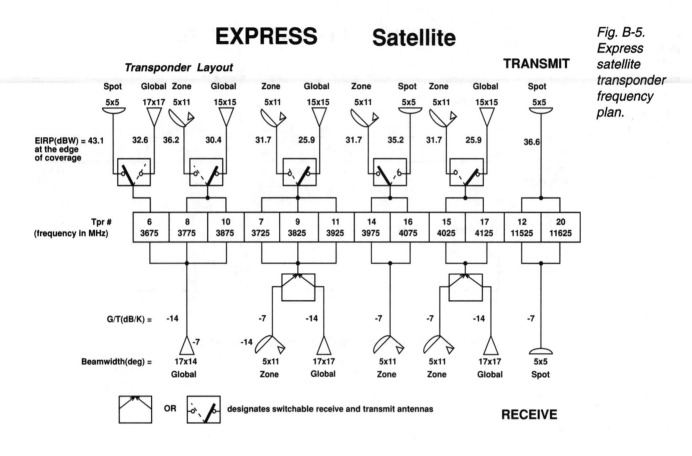

Fig. B-5.
Express
satellite
transponder
frequency
plan.

Transponder Layout

TRANSMIT

RECEIVE

*Fig. B-6.
INTELSAT
805 satellite
transponder
frequency
plan.*

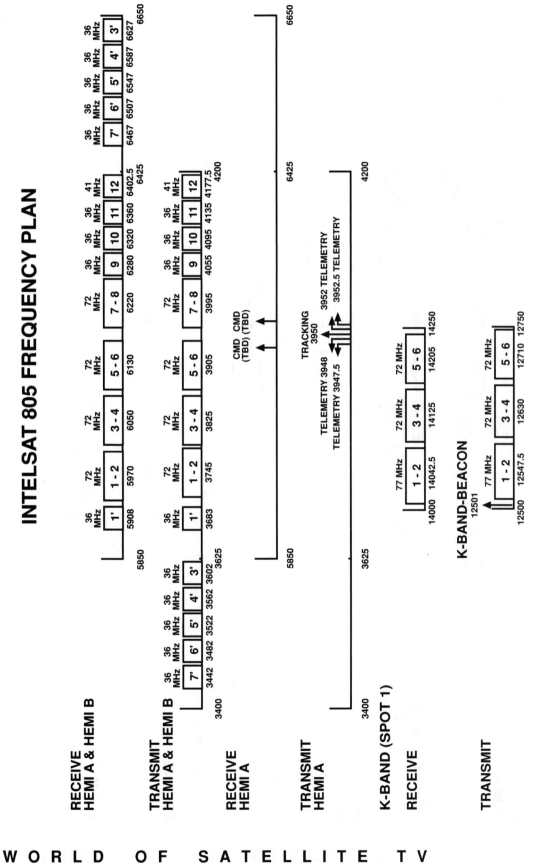

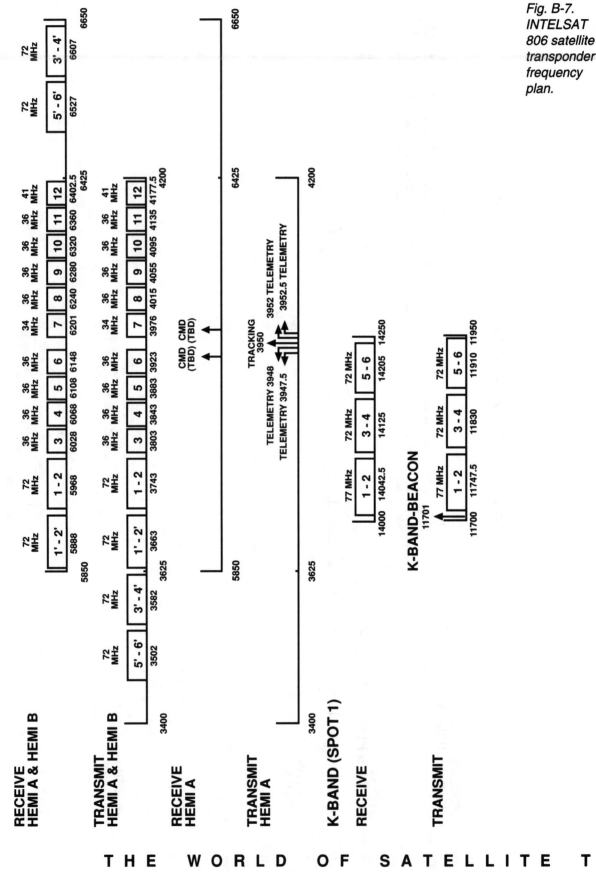

Fig. B-7. INTELSAT 806 satellite transponder frequency plan.

Fig. B-8. Nahuel 1A Ku-band transponder frequency plan.

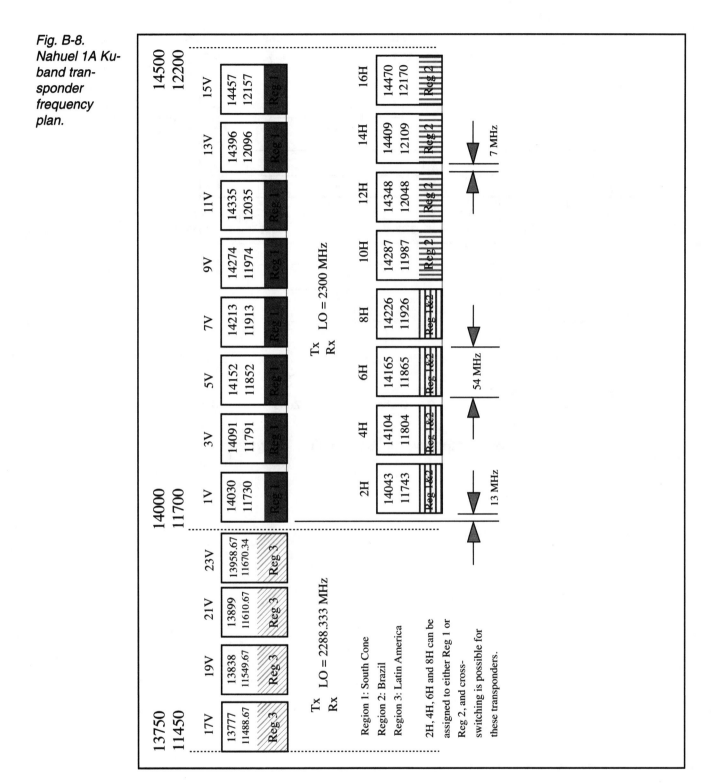

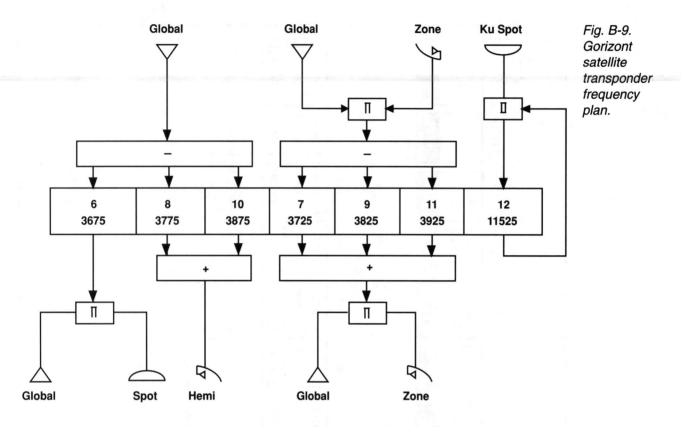

Fig. B-9. Gorizont satellite transponder frequency plan.

Gorizont Antenna / Transponder Connection Diagram

Appendix C: A Glossary of Terms

A

Algorithm-A mathematical process used to encode video, audio and data signals for encryption and/or compression purposes.

AM-Amplitude modulation causes the baseband signal to vary the frequency of the main transmission carrier wave.

Analog-A signal waveform that continuously varies in both intensity and frequency, potentially generating an unlimited number of possible values. All of the signals that naturally occur in our environment, such as the human voice, birds singing, or the sound of a waterfall, are all analog signals.

Antenna-Any manufactured device constructed and used to collect or radiate electromagnetic waves of energy for the purpose of gathering or disseminating information.

AOR-The Atlantic Ocean Region as defined by international satellite operators such as INTELSAT and PanAmSat.

Aperture-The effective capture area of a communications antenna that is capable of intercepting the incoming signal.

Arc zenith-The highest point in the geostationary arc aligned to the true north/south line crossing through the site location.

Ariane 4-A satellite launch vehicle developed under the auspices of the European Space Agency.

Artifacts-Visual impairments in the video signal due to the limitations of the TV transmission standard in use.

ASCII-The American Standard Code for Information Interchange is a worldwide standard for text coding that is used and understood by all computer systems regardless of differences in operating systems.

Aspect ratio-The picture width to picture height ratio of a TV screen's display area.

ATS-Applications Technology Satellite. Early series of NASA experimental satellites that laid the groundwork for the modern satellite communications era.

Attenuation-A measurement of signal loss within a transmission system or component that is expressed in decibels (dB).

Axis of symmetry-The bisecting line which divides the paraboloid antenna into two symmetrical halves.

AZ-The azimuth angle (see azimuth).

Azimuth-The horizontal angle between the antenna's main beam and the meridian plane of the site location.

B

Backhaul-A communications link connecting a remote communications station to a local switching center or network control center.

Band-A specific range of frequencies in the electromagnetic spectrum.

Bandwidth-A range of frequencies occupied by a communications signal or passed through a transmission channel. Also a measure of the information-carrying capacity of a communications channel. A communications signal that requires a bandwidth greater than 20 kHz is called a "broadband" signal, while those with a bandwidth smaller than 20 kHz are called narrowband signals.

Baseband video-For satellite TV applications, this is a spectrum of frequencies from 0 to 8.5 MHz that contains the video and audio information, including any auxiliary subscarriers, prior to demodulation by the receiver or IRD.

Beamwidth-The aperture window or acceptance angle of any antenna, which is typically measured between -3 dB half power points for paraboloid antennas.

BER (Bit Error Rate)-A measurement of the accuracy of digital decoding in a receiver or IRD, typically expressed in one of two forms of mathematical notation, such as 1×10^{-3} or 1E-3. In digital DTH systems, the threshold point of the IRD is typically defined as a specific BER.

Bit-A binary digit, either 0 for "off" or 1 for "on."

Bit Rate-The speed at which a digital transmission occurs, measured in units of bits per second (b/s), a thousand bits per second (kb/s), or a million bits per second (Mb/s).

Blanking Interval(s)-The unseen horizontal and vertical portions of the video signal that contain syncronizing information for the TV set, and, in the case of analog encrypted signals, also contain conditional access and/or digital audio signals.

Block-A predetermined group or string of binary digits. Also an 8 x 8 pixel array that is converted from a spatial domain to a frequency domain in a digital video compression system. An MPEG-2 16 x 16 pixel array is also called a macroblock.

Block Coding-A digital coding system in which the encoder looks only at the bits contained within the current block of data.

Block Downconversion (See also LNB and LNF)-The use of a fixed-frequency local oscillator, which heterodynes with the incoming microwave frequency spectrum to convert an entire satellite band to a lower band of intermediate frequences for transmission to the receiver or IRD for demodulation.

Bouquet-A unified group of digitally compressed TV, audio, and data signals.

BSS (Broadcast Satellite Service)-The official term used to refer to high-power DBS satellites and services using specific frequencies, polarizations, and orbital locations that have been assigned by the ITU and/or its sanctioned regional and world radio administrative conferences.

b/s-bits per second.

Byte-A digital "word" consisting of eight bits. Used to describe information storage and retrieval of data in computer systems.

C

Carrier-The specific radio frequency that is modulated by a baseband signal to relay information through a communications system.

Cassegrain-A type of satellite antenna that employs a convex hyperboloidal subreflector and paraboloid main reflector to reflect signals to a focus located within the central portion of the dish.

C-band-Typically used to refer to a microwave frequency range that extends from 3.4 to 4.8 GHz and is used to receive or downlink signals from communications satellites.

Channel-For terrestrial TV and radio broadcast systems, this is a one-way communications link from the broadcaster to the viewer or listener.

Chrominance (Chroma)-The portion of the baseband video signal that contains the color information.

Clarke Orbit-(See geostationary arc.)

CNR (or C/N)-The Carrier to Noise ratio produced by the satellite receiving system, which provides the means for quantifying the performance of the system. The threshold of an analog receiver or IRD is typically expressed as a value of C/N in dB.

Collocation-The location of multiple satellites at approximately the same geostationary orbital location (usually with a minimum separation of 0.2 degrees between each satellite in the "constellation."

Color burst-A synchronizing component of the video waveform that serves as the frequency and phase reference for the color, or chorminance, information.

Color Subcarrier-For NTSC systems, a subcarrier centered on a frequency of 3.579545 MHz (in reference to the main carrier frequency) that relays the color information component of the TV signal.

Common Carrier-A U.S. telecommunications service provider that operates under rates and rules established by the Federal Communications Commission.

Composite Baseband-The raw signal output of a satellite receiver that contains all components of the baseband video signal. This unfiltered output signal is normally used to connect to a stand-alone decoder for reception of an encrypted satellite TV service.

Composite Video-The entire video signal, which includes the synchronizing, luminance, and chrominance information but does not contain any audio or data subcarriers that may be present in the demodulated baseband signal.

CONUS-contiguous United States (not including Alaska or Hawaii).

Compression-Removal of information from a communications signal in order to reduce the amount of transmission bandwidth required. The information removed is either nonessential to human perception or can be restored at the receive end.

Concatenation-The use of two coding systems, whereby the output of the inner encoder falls or cascades into the outer encoder. Used by all MPEG-2 DVB-compliant DTH services.

Conditional access (CA)-The authorization system and/or data which allows a decoder to access and process an encrypted communications signal.

Convolutional coding-A digital coding system which incorporates memory so that the encoder can look at both the previous and current blocks of data.

CRT-Cathode Ray Tube. The standard image display unit used in all TV sets.

CTS (Hermes)-The Communications Technology Satellite that provided the earliest experimental high-power DTH transmissions in the Ku-band frequency range.

Cycle-A sine wave with 360 degrees of revolution. (See also Hertz.)

Cycles per second-The measurement unit of frequency, also called a Hertz after the 19th century discoverer of "Hertzian waves." (See also Hertz.)

D

dB (decibel)-The logarithmic expression of the ratio between two numbers, with 3 dB represents a doubling factor, 10 dB a times 10 (x 10) multiplication factor, 20 dB times 100 (x 100), and so on.

dBi-The gain in decibels of a communications antenna in reference to an isotropic source.

dBm-An expression of dB power relative to one milliwatt.

DBS-Direct Broadcast Satellite. (Often used interchangeably with the more formal term BSS, for Broadcast Satellite Service.)

dBW- The expression of satellite power, called the effective isotropic radiated power (EIRP), in decibels relative to one watt of power.

DCT (discrete cosign transform)-In an MPEG-2 compression system, the DCT is used to convert blocks from a spatial domain to an equivalent set of DCT coefficients that are expressed in a frequency domain.

Decoder-The processing unit at the receive end of a communications link. Information that has been coded at the transmit end, either to improve signal quality and/or to encrypt the information, is restored to its original state. (See also IRD.)

Decoding Margin-In digital systems, the extent by which the Eb/No exceeds the BER that equates to the receiving system's digital threshold. (See also Eb/No.)

Deviation-In FM communications systems, the extent to which the modulating baseband signal shifts the main carrier frequency from its nominal assigned value.

Discriminator-A type of FM demodulator used in analog satellite TV receivers.

Domsat-A domestic communications satellite.

Downconversion-The translation of a frequency or entire block of frequencies to a lower intermediary frequency (IF) spectrum. (See also block downconversion.)

Downlink-The space-to-Earth side of a two-way satellite communications link. Also may refer to the system used to receive the downlink signal.

DRO-The Dielectric Resonant Oscillator used in most LNBs and LNFs to provide a relative measure of stability for the block downconversion of the microwave signal to an intermediate frequency (IF) band.

DTH (Direct To Home)-Generally used to describe any satellite TV or audio transmission that is expressly intended for direct broadcast reception at individual homes.

DTV-A new digital TV standard for the USA that already has been adopted by the FCC.

DVB (Digital Video Broadcasting)-A digital video compression standard for broadcasters that incorporates the MPEG-2 specification as a subset. Communications systems that use the DVB standard are said to be DVB-compliant. (See also MPEG.)

DVB Group-The European organization that cooperatively established the DVB standard.

Eb/No-In digital transmission systems, the Energy-per-Bit-to-Noise-power-density ratio is the digital equivalent to the C/N or C/NR used to evaluate analog system performance.

Eclipse-The time period when any satellite crosses into the shadow of the Earth or moon, during which the spacecraft must operate from its storage batteries rather than from the power normally supplied by the satellite's photovoltaic solar cells.

ECM-Electronic Counter Measure. A security enhancement employed by a program service operator to thwart signal piracy.

Edge of coverage (EOC) -The satellite operator's defined service area for a satellite coverage beam or "footprint," which is typically 3 or more dB down from the peak signal level to be found at beam center.

EIRP-Expressed in dBW, the Effective Isotropic Radiated Power numbers indicate the signal levels for each contour on a satellite coverage map. The EIRP represents the combination of the power in watts produced by the satellite transponder's amplifier and the gain of the satellite's transmission antenna, minus any internal waveguide or multiplexer losses between the two components.

EL/AZ (or EL over AZ)-A type of paraboloid antenna mount that provides for independent steering of the azimuth and elevation coordinates.

E

EIRP (or e.i.r.p.)-The effective isotropic radiated power of a satellite signal represents the combination of the transponder amplifier power output in watts plus the gain in dB of the satellite's transmitting antenna, hence the measurement term dBW.

Electromagnetic Spectrum-The entire range of frequencies from very long waves to visible light, all of which radiate outward from the point of origin at the speed of light.

Elevation-The angle between the antenna's main beam and the horizontal plane.

Encoder-An electronic device that converts information to an alternate format that requires a matching decoder to convert the encoded message back to its original uncoded state. Encoding may be used to transmit data in a more compact form, to encrypt the data to prevent unauthorized reception, or both.

Encryption-The encoding of any communications signal using secure electronic keys or algorithms to prevent unauthorized reception of the signal.

EPG-Electronic Program Guide.

f/D-The focal length to Diameter ratio of a paraboloid reflector.

Faraday Rotation-The rotation of linear polarization satellite signals that pass through the Earth's ionosphere during peak periods of sunspot activity.

FCC-The Federal Communications Commission is the U.S. government agency established in 1934 to regulate all electronic forms of domestic communication.

FEC (Forward Error Correction)-Used by many digital transmission systems, FEC is a coding technique that supplies the receiver with data that can be used to recover essential information that may be lost due to link noise.

F

Feedhorn-The device located at the focal point of a parabolic antenna that is said to illuminate the reflector in such a way that the maximum amount of signal and lowest possible level of noise is retrieved. (See also LNF.)

Field-One half of a complete vertical scan of a TV image consisting of either the even-numbered or odd-numbered lines. The two fields are combined or interlaced to make each complete video image or frame.

Footprint-The coverage area or beam pattern produced by a communications satellite. Also refers to any map of the satellite coverage zone that contains contours showing the EIRP or equivalent antenna diameters required to receive the signal at locations within the general vicinity of each contour.

Frame-One complete vertical scan of a video image, consisting of two interlaced fields.

Frequency-The number of times that an alternating current goes through a single 360-degree revolution or cycle in one second of time. One cycle per second is also called a hertz; 1,000 cycles per second a kilohertz; 1,000,000 cycles per second a megahertz; and 1,000,000,000 cycles per second a gigahertz.

Frequency reuse-Multiple use of an available communications spectrum through the application of orthogonal polarizations and/or multiple coverage beams that are spacially isolated from one another.

FSS (Fixed Satellite Service)-The official term used to refer to satellites and satellite services using specific frequencies assigned by the ITU in cooperation with regional and world radio administrative conferences that originally were not intended for direct reception by the general public.

G

Geostationary arc (or geostationary orbit)-A unique orbit in the plane of the Earth's equator at a distance of some 22,300 miles in which a satellite can remain essentially stationary relative to locations on the Earth below.

GHz (or gHz)-A gigahertz is a unit of frequency that is equal to 1,000 MHz, or one billion cycles per second.

Global Beam-A type of satellite coverage beam used by INTELSAT and others to illuminate that portion of the Earth's surface which is visible from the satellite's orbital location.

G/T (G over T)-The antenna gain to system noise temperature ratio, expressed in dB/K, which is considered the ultimate figure of merit for evaluating receiving system performance.

H

Half Transponder-A transmission method used by INTELSAT and other satellite operators whereby two analog TV signals are transmitted simultaneously through a single satellite transponder. This is accomplished by reducing the FM deviation and power allocation for each TV signal.

HDTV-High Definition Television.

Hemispheric Beam-A type of satellite coverage beam used by INTELSAT and others to illuminate approximately one-half of the Earth's surface that is visible from the satellite's orbital location.

Hertz-The standard unit of frequency named after the 19th century scientist who first experimented with the transmission and reception of radio waves, also known as Hertzian waves. One hertz equals one cycle per second.

High Power Satellite-Any communications satellite carrying power output amplifers that exceed 100 watts of power.

I

IF (Intermediate Frequency)-Often used to refer to the signal output of an LNB or LNF or the input of a satellite receiver or IRD.

Indoor Unit-A term used by some digital DTH operators to refer to the IRD. Also called a receiver or a set-top box.

INTELSAT-The International Telecommunications Satellite Organization is a global satellite cooperative that operates more than twenty-five communications satellite in geostationary orbit.

IOR-Indian Ocean Region as defined by international satellite operators such as INTELSAT and PanAmSat.

IRD-Integrated receiver/decoder. Also called the indoor unit, receiver or the set-top box.

ISDN-The Integrated Services Digital Network is a high-quality digital tele-communications network that allows subscribers to access numerous types of communication services through a single network connection.

Isotropic antenna (or isotropic source) - A hypothetical antenna that would radiate (or receive) energy equally well in all directions at once. Used as a reference model for making antenna gain measurements.

ITU-International Telecommunication Union.

J

JPEG-Digital compression standard used for computer graphics and an early precursor to the MPEG compression standard. M-JPEG was one derivative of this standard used early on to compress moving pictures.

K

Ka-Band-Generally used to refer to the downlink frequency spectrum extending from 18 to 22 GHz.

Kelvin (K)-The unit used in noise measurement for expressing absolute temperature.

Ku-Band-Generally used to refer to the downlink frequency spectrum from 10.7 to 12.7 GHz.

L

Latitude-The distance in (0 to 90) degrees from the Earth's equator to points north or south.

L-band-Generally used to refer to non-geostationary LEO satellites that downlink signal in the region of 1.6 GHz.

LEO-Low Earth Orbit.

LHCP-Left-Hand Circular Polarization.

Line Rotation-Analog video encryption technique. Also called the "cut and rotate" method.

LNB-Low Noise Block Downconverter. (See also block downconversion.)

LNF-A combination of a feedhorn, automatic polarization device, and LNB in a single sealed package. (See also block downconversion and feedhorn.)

Longitude-The distance in degrees (either 0 to 360 degrees east longitude, or 0 to 180 east and 0 to 180 degrees west longitude) from one meridian to any other. INTELSAT (as well as satellite operators located in the eastern hemisphere) normally refers to the orbital locations for their satellites in terms of degrees east longitude, while all other satellite operators in the western hemisphere refer to the orbital locations for their satellites in terms of degrees west longitude. An INTELSAT satellite located at 307 degress east longitude would have an equivalent orbital location of 53 degrees west longitude.

Look Angle-The angle at which the antenna tilts up (in reference to the horizon) to receive signals from a communications satellite.

Low Power Satellite-Typically used to describe any satellite carrying power amplifers that produce less than 30 watts of output power.

Luminance-The brightness componant of the video signal that contains the lighting and shading information.

M

Mb/s-Megabits per second (also expressed as Mbits/s).

Medium Power Satellite-Typically used to describe any satellite carrying amplifiers that produce 30 to 100 watts of output power.

Megabyte (Mbyte)-A unit of measurement commonly used to describe the information storage capacity of computer systems, which is equal to 1024 x 1024 bytes, or 8 Megabits.

MegaSymbol-(See symbol.)

Meridians-Imaginary lines circling from pole to pole which cross each of the equator's 360 degrees.

Modem (Modulator/demodulator)-A communications device used for converting digital data to audio tones for low to medium speed transmission of information over a telephone circuit or other non-digital transmission medium.
Modulation-The attachment of information to a radio frequency carrier wave through the deviation of one or more parameters of the carrier wave.

Typical modulation methods vary the amplitude, frequency, or phase of the carrier wave in accordance with the instantaneous amplitude and/or frequency deviations of the modulating information.

Modulator, RF- A device contained within the satellite receiver or IRD that remodulates the satellite video and audio signals onto an RF carrier that corresponds to a standard VHF or UHF TV channel frequency to deliver TV signals to one or more TV sets in the home.

MPEG-Motion Pictures Experts Group. Also used to refer to the standards developed by this organization. The MPEG-1 compression system was developed for all progressive (non-interlace) scan sources of multimedia such as text, graphics and film. The MPEG-2 compression system was developed for all interlace scanning sources of media such as video recordings and broadcast TV.

Multiplex (Digital)-The combination of multiple video, audio and data signals into a single unified digital bitstream.

N

NASA-National Aeronautics and Space Administration.

Noise-Any unwanted signals present in a communications medium that can degrade the reception of the intended signal. The two types of noise that typically affect satellite TV reception are thermal noise generated by molecular motion in all matter and impulse noise generated by manufactured devices such as internal combustion engines and microwave ovens.

Noise Figure-The measurement, expressed in dB, of the internal noise contribution of a Ku-band KNB or LNF.

Noise Temperature-The measurement, expressed in degrees Kelvin (K), of the internal noise contribution of a C-band LNB, or LNF.

NTSC-National Television Standards Committee. Also used to refer to the video standard presently used in the United States and other countries around the world. The characteristics of the NTSC standard are as follows: 525-lines per frame transmitted at a frame rate of thirty frames per second, with each frame consisting of two interlaced fields that alternate at a field rate of sixty fields per second. The horizontal resolution of NTSC is 768 picture elements, or pixels, per line. The NTSC video signal has a broadcast bandwidth of 4 Mhz.

O

Offset (fed) Antenna-An antenna design with a reflector that only forms a part of a complete paraboloid of revolution, such that the focal point will be located out of the path of the incoming satellite signal. The offset angle is the deviation in degrees from the axis of symmetry of a paraboloid used to produce an offset-fed antenna.

Orthogonal-Mutually at right angles. (Used to describe the isolation between either horizontal and vertical polarization, or right-hand and left-hand circular polarization.)

Outdoor Unit-The antenna and LNF (or LNB and feedhorn).

P

PAL-Phase Alternation (by) Line. Color TV standard developed in Germany.

PRBS-Pseudo Random Binary Sequence (generator) used to generate the electronic keys used to encrypt satellite TV broadcasts.

PID-The Picture Identification number used to identify the location of an individual video service within a DVB-compliant digital bitstream.

Pixel-Picture element.

Planar Array-A type of flat satellite antenna developed in Japan that is composed of a gridwork of tiny resonant elements.

Polar Mount (Modified)-Antenna mount mechanism that permits the steering of the dish to track the available satellite in geostationary orbit by the rotation of the reflector about a single axis. Radio astromomers use a classical polar mount where the mount axis is precisely parallel to Earth's polar axis. The modified polar mount used for large dish DTH applications incorporates a declination offset to compensate for the relative proximity of the satellites as opposed to the stars which are light years away.

Polarization-The direction or rotation sense of an electromagnetic wave.

POR-Pacific Ocean Region as defined by international satellite operators such as INTELSAT and PanAmSat.

Q

QAM-Quadrature Amplitude Modulation is the modulation system used to transmit digital TV signals over cable TV systems or to broadcast DTV signals terrestrially.

QPSK-Quadrature Phase Shift Keying (or Quaternary Phase Shift Keying) is the modulation system used to transmit digital TV signals over satellites.

R

Rain Outage-The loss of signal that commonly occurs at frequencies above 10 Ghz due to signal absorption and an increased thermal noise caused by heavy rainfall.

RARC-Regional Administrative Radio Conference.

Reed Solomon-An FEC coding technique used by all DVB-compliant satellite TV transmission systems.

RF-Radio Frequency.

RF Modulator (See Modulator, RF.)

RGB (Red, Green, Blue)-The three primary color components in a TV system. (See also YUV.)

RHCP-Right-Hand Circular Polarization.

S

S-band-Generally used to refer to satellites or satellite signals which downlink in the 2.5 to 2.6 GHz frequency spectrum.

SECAM (Sequence Couleur avec Memoir)-A 625-line color TV standard developed in France.

Sidelobe-The off-axis response of a satellite dish.

Skew-the difference (real or apparent) in polarization angle between two or more satellties using linear polarization.

SMATV-Satellite Master Antenna Television.

S/NR (Signal-to-Noise Ratio)-A standard measurement, in dB, of the video signal level that is used to evaluate video performance.

Solar Eclipse-An event that predictably occurs twice each year during the two-week period surrounding the equinoxes when the Earth will shadow the satellite's solar array from receiving energy from the sun, forcing the spacecraft to draw power from its storage batteries. (See also solar outage.)

Solar Outage-An event that predictably occurs twice each year during the two-week period surrounding the equinoxes when the sun will pass directly through the satellite antenna's main beam. The Sun has a noise temperature of 10,000 K. This causes a loss of signal, which normally lasts for just a few minutes.

Sparklies-A non-technical term used by some to describe the impulse noise spikes that appear as comet-tailed black or white spots in an analog satellite TV picture when the receiver or IRD is operating near or below its threshold rating.

Spherical-A multiple beam antenna which uses the simple geometry of a sphere.

Spot Beam-A tightly focused satellite coverage beam, usually with either a circular or an elliptical shape.

Stationkeeping-Ground controlled adjustments to the orbit of a geostationary satellite that assist the spacecraft in maintaining its assigned location over the Earth's equator. As these maneuvers require the periodic expenditure of fuel, the prime determinant of spacecraft life is the amount of stationkeeping fuel that it carries into orbit.

Subcarrier-An information carrying wave that is attached to the main carrier of a broadband transmission, the frequency of which is referenced to the frequency of the main carrier. Within the baseband video signal, subcarriers can be used to convey color information, TV sound, auxiliary audio, and data signals.

Subsatellite point-The spot on the Earth's equator over which the geostationary satellite is located.

T

TDRSS-The Tracking and Data Relay Satellite System used by NASA to coordinate shuttle launches and by U.S. satellite operator Columbia Communications Corporation for international service.

Teletext-An on-screen text information service that is transmitted in a digital format within the TV signal's vertical blanking interval.

Threshold-In an analog satellite TV system, the value of CNR at which the relationship between incoming signal CNR and the demodulated video signal SNR becomes nonlinear. In a digital DTH system, the digital threshold is defined as the bit error rate (BER) at which the IRD ceases to function.

Threshold Extension-Techniques employed by some analog satellite TV receivers and IRDs to reduce the CNR threshold.

TI (Terrestrial Interference)- Interference to satellite reception caused by ground-based microwave transmitting stations.

Transmodulator-The electronic device used in digital cable head ends to convert DVB satellite signals using QPSK modulation to QAM modulated signals for transport through a digital cable TV distribution system.

Transponder-The combintation of an uplink reciever and a downlink transmitter which acts as a repeater of one or more satellite signals.

TVRO-Television Receive Only.

U

UHF-A frequency spectrum ranging from 300 MHz through 3 GHz.
Uplink-The space-to-Earth signal pathway of a satellite transmission.

V

VBI-The Vertical-Blanking Interval of a TV signal.

VHF-A frequency spectrum ranging from 30 MHz through 300 MHz.

W

WARC-World Administrative Radio Conference. (Recently shortened to WRC for World Radio Conference.)

X

X-band-Generally used to refer to military communications satellites and signals operating in the 7 to 8 GHz frequency range.

Y

YUV-An alternative method of representing the components of a video signal with Y representing the luminance signal and U and V the normalized versions of the blue and red color-difference signals, B-Y and R-Y. In digital TV systems, the luminance and chrominance information is represented by the Cr and Cb components.

Index